AF846060

John Curtis and the pioneering of pest control

The Great Innovators

G. E. Fussell	*Jethro Tull: his influence on mechanized agriculture*
Peter Hay	*Brunel: his achievements in the transport revolution*
Raymond Watkinson	*Bewick and the rise of illustrated journalism*
George Ordish	*John Curtis and the pioneering of pest control*

John Curtis and the pioneering of pest control

George Ordish

OSPREY

Published in 1974 by
Osprey Publishing Ltd
707 Oxford Road, Reading, Berkshire RG3 1JB
© Copyright 1974 George Ordish
All rights reserved

ISBN 0 85045 159 0

To Meliora Dockery

WITHDRAWN FROM
UNIVERSITY OF LEICESTER
LIBRARY

LEICESTER UNIVERSITY LIBRARY

RB m000376

CUR | ORD
449178
26 4 74

Text set in 11/12 pt. Monotype Baskerville, printed by letterpress,
and bound in Great Britain at The Pitman Press, Bath

Preface

I am much indebted to many people for help with this book, particularly to the Royal Entomological Society and its Registrar, Miss E. Evans, for permission to peruse and quote from Curtis's letters and diary in their possession, to Mr Th. O'Grady of the Linnean Society, to Mr E. R. C. Tasker of the Royal Agricultural Society, to Dr Gwynne Vevers of the Zoological Society, London, to Miss J. M. Kennedy, the Norwich archivist and her staff and Dr Rajnai of the Castle Museum, Norwich, to Mr D. M. Hammond of the Natural History Museum, London, and the archivists of numerous banks (the Bank of England; Miss J. M. R. Campbell, National Westminster; S. R. Maycroft, Midland; R. O. Richards, Lloyds; Miss K. Byron, Barclays) and to Mrs S. Feakin, and Messrs T. B. Norgate, Lt Col W. Ruel and E. J. Troon for help with Curtis's family and residences.

Finally I would like to thank my wife for reading the manuscript and for much help with it.

John Curtis was a remarkable man and it is hoped this book will throw a little more light on his life, achievement and importance in crop pest control.

GEORGE ORDISH

St Albans.
February 1973.

Contents

Illustrations

(appearing between pages 56 and 57)

1 Norwich

John Curtis was a remarkable man in two respects. In the first place he was a very able scientist and observer, but still more remarkable was the fact that he was one of the first *professional* scientists, those men who live by practising science. Curtis was born in 1791 and was thus a young man in the early nineteenth century. Professional scientists, apart from the doctors (and then medicine was more an art than a science), were very rare and almost unknown among biologists. Scientists tended to be wealthy men, the children of wealthy men, or at any rate of those with a good middle-class competence, who took to science as an intellectual interest, or they might be comparatively poor but under the patronage of a wealthy man or institution, such as a university or a government department. For instance, Sir Humphrey Davy (1778–1829), who did so much for chemistry, mining and agriculture, was a property-owning Cornish baronet. Baron Cuvier (1769–1832), the naturalist, who 'having a decided partiality for natural history, devoted his leisure hours to the pursuit of that interesting science', was born into the court circle of the Duke of Wurtemberg and caught the eye of Napoleon, hence his imperial barony. Charles Darwin (1809–82) was the son of a wealthy Shrewsbury doctor, whose money seems to have come more from property appreciation and Wedgewood pottery than the exploitation of the practice's patients. The Reverend M. J. Berkeley (1803–89), who invented the science of 'vegetable pathology' and did much to control the diseases of crops, received a university education and became a curate at Apethorpe, Northamptonshire and, whilst this was no sinecure, he does seem to have had an understanding rector and thus was able to devote much time to his hobby. The entomologist James F. Stephens (1792–1852), about whom we shall have more to say later, was the son of an admiral, obtained a position in the Admiralty Office, and was transferred to the British Museum, on full pay, where he arranged the collection of insects and wrote articles and books for the Press – an example of official patronage.

John Curtis, and Michael Faraday too, were both poor men (and sons of poor men) who took up science as a career and succeeded in it, though it was a hard road to follow. They were both born in September 1791 and, having survived to manhood, had a great struggle to live from the exercise of their respective professions, though both were fortunate in their early patrons, particularly Faraday, who was greatly helped by Sir Humphrey Davy and the Royal Institution. The life of the patronized can be hard and frustrating and is neatly epitomized in Dr Samuel Johnson's (1709–84) Dictionary.[38] The work was written under the patronage of Lord Chesterfield, who treated the learned doctor in a cavalier fashion, until, finding the Dictionary a success, the former tried both to claim most of the credit and to make amends by writing an article in *The World*, an action which but further annoyed Johnson. In the Dictionary, under 'Patron', we find:

1. One who countenances, supports or protects. Commonly a wretch who supports with insolence, and is paid with flattery.

And later we have:

To Patronage. (v.2.) (from the noun). To patronise, to protect. A bad word.

Some small revenge for years of slight!

Today the complete opposite is the case. The modern world belongs to the scientist who, working for governments or large industrial concerns, is well paid and generally leads an interesting life as he 'unravels the secrets of nature'. Galbraith[33] even maintains that as modern technology is so complex and thus needs teams of scientists to run it and its complex plant, these men then tend to produce what it suits them to produce and not necessarily what the consumer wants, except, of course, that modern scientific psychologists will persuade the consumer that what it suits the firm's technology to produce is what he, the consumer, wants. It is done by means of advertising.

Curtis was not as fortunate as Faraday in that at a time when biology was seething with new ideas which were to terminate with the *Origin of species by means of natural selection*[23], he had to struggle to support himself and his family. That he survived and did important work was remarkable.

The evidence for the early life of John Curtis is scarce and what there is is discussed in Appendix 1. Briefly the conclusions we can draw from it are: John Curtis was born in the lovely town of Norwich on 3 September 1791 of a respectable artisan family. His father was Charles Morgan Curtis, an engraver and sign painter and possibly a verger. He may have owned a house in the Cathedral Close, in the parish of St Mary in the Marsh, the cathedral serving also as the parish church. The cathedral was also known as the Church of the Undivided Trinity. A house in the Close was entered in the Land Tax returns (1789–95) as belonging to Morgan Curtis, although occupied at the time by a Robert Osborn, at first for a rent of £3 a year (1789 and '90), then later at £9 (1791–3) which unaccountably fell back to £3 a year (1794–6).

The Christian name of Charles's wife was Frances, but two generations of searchers have failed as yet to discover her maiden name. While seeking vainly for that information among the records I myself came across a discouraging letter written in 1901 by C. W. Dale, the son of John Curtis's great friend and supporter, J. C. Dale. It was in reply to J. H. Hooperhoe, who was then writing an account of Curtis's life for the *Dictionary of National Biography*,[26] and had asked for details that might help him in a task that he was finding difficult. The letter runs as follows:

> The family was possibly not a Norwich one and that may account for your difficulty in getting information. I do not know what the names of Curtis's mother or wife were. The family was not high enough to have a genealogical tree.

Seventy-one years later it is still difficult, but slightly less so because the records at Norwich are now so well arranged. Curtis is a common name in the area.

It would appear that the first child of Charles and Frances was a daughter, Frances, born about 1789, who, according to the record of a memorial to her, died, aged three, in October 1792.

Their eldest son was born in 1790 and named Charles Morgan after his father. Then came John, the subject of this book, born on 3 September 1791. Between that date and 1795 there were presumably at least two more children, because John Curtis in later years talks of a sister living in Welbeck Street, London, and the memorial mentioned above, after commemorating the child Frances, continues 'also a son aged fifteen months'.

As happened so frequently in those days of high infant mortality, little Charles Morgan, like the first child, died very young, at the age of five. Not long afterwards, on 15 October 1796, another son was born and called after his recently deceased brother, as well as his father, Charles Morgan. This Charles Morgan Curtis junior grew up to be an engraver like the other two male members of his family.

The father of this family died on 4 January 1796, soon after the birth of the youngest child, leaving his widow and children in difficult circumstances. The house in the Lower Close was sold, became the junior verger's house and was occupied by the Reverend Dr Forster. Frances Curtis, the widow, became a 'cultivator of flowers' and thus helped support her family. Although she had a hard time she was able to send at least the two boys to school, probably the Norwich Grammar School.*

In due course the house in the Close became the property of Deacon Clark and was occupied from 1807 to at least 1817, and probably until 1822, by 'Mrs Curtis and others'. On 30 November Mrs Curtis died, aged sixty-three. She was thus thirty years old when the first of her many children (at least six) was born, a considerable age in those days to bear a first child.

The boy, John, became interested in natural history at an early age, particularly in insects, after watching the garden tiger moth through all its stages of development. We can picture him fascinated by the black and white woolly caterpillars of this moth, probably feeding some of them on nettles or lettuce in a cage, watching them pupate and then patiently waiting for the early summer when the striking looking adults would appear – the forewings brown with interlaced white bands and spots and the hind wings red or orange with black spots sometimes surrounded with a yellow margin.

In addition to the fascination of watching these creatures, the young boy also found it profitable, because there are accounts of his selling an adult lobster moth (*Staurops fagi*), which he had found as a caterpillar on the lime trees in the Cathedral Close and then reared, for £5 – an enormous sum for those days, particularly as it is the caterpillar that is so striking, the moth itself being a dull, unremarkable, brownish creature. The caterpillar has elevated conical humps on the rear side of the back,

* The Grammar School records were lost in a fire in 1855.

the ring divisions of the body are well defined and the true legs are long and jointed, in fact it distinctly resembles a tiny lobster. Presumably some rich man in the neighbourhood very much needed one for his collection, or alternatively some well-wisher wanted to help the widow and her family in a kindly unobtrusive way.

The child used to study the plants and insects of ponds and marshes with Richard Walker, a local naturalist, and was much influenced by him. There is an account of the two young naturalists being particularly attracted by the movements of the water boatman beetles. Walker eventually became a Fellow of Magdalen College, Oxford and wrote *Flora Oxoniensis*.[84] This devotion to the task is said to have led to a severe attack of rheumatic fever which almost killed young Curtis. We might think today that it was more likely to have been malaria.

On his recovery his mother was able to send him back to school at Norwich where he became friendly with a fellow pupil – Henry Browne – whose mother had a collection of British Lepidoptera. Young Curtis was allowed to examine them. This further increased his knowledge and interest in the subject. He must have been a likeable child; he became acquainted with and retained the interest of Dr (later Sir James) Edward Smith and a Mr Hooker, the father of the famous Sir W. J. Hooker, the botanist and director of Kew Gardens. The young Hooker helped Curtis in the naming of insects, whose only guide up to this point was Berkenhout's *Synopsis of British Natural History*,[4] a rather weak reed, and there is a story of their capture of both adults and larvae of the rare swallow tail butterfly in the Horning marshes. The boy early on showed an appreciation of line and colour which persisted the whole of his life and stood him in good stead. He liked to colour engravings, odd ones left by his father and those in his precious Berkenhout; he also started to copy Donovan's *British Insects*,[29] which he probably borrowed from Walker or Hooker. It was an expensive work in several volumes and Curtis is unlikely to have owned a copy.

In spite of his fascination with plants and insects their study was no profession for a poor, young boy and he was placed at the age of sixteen (1807) in a Norwich lawyer's office as a writing clerk. Young John was then shut up in a dusty office, seated at a high stool and desk, endlessly copying letters, wills and accounts when he longed to be out collecting, naming and studying plants and insects or at an easel drawing the charming East Anglian landscape, for the young man had become one of 'Old Crome's' circle of enthusiastic students. He can, however, have had very little time for such pursuits for there was no nonsense about half-days on Saturday in those times, and generally the Sabbath could not be broken, though painting, botanizing and collecting insects might have been considered innocent enough.

Curtis stayed with the lawyer some four years, most of which he must just have tolerated for the sake of his mother and the rest of the family. During this period he made friends with Simon Wilkin, from the same neighbourhood. Wilkin was a rich young man interested in natural history and insect collecting.

The budding naturalist (twenty years old) was now launched on his career of entomologist and engraver. He met a number of scientific clergymen, including the Reverend William Kirby, who, with his friend the Reverend William Spence, was to popularize the science with their famous book *An Introduction to Entomology*,[41] and the Reverend John Burrell and others. When he reached his majority, Wilkin went to live on his property at Costessey and took the young Curtis with him, there to arrange and catalogue his collection of insects. The life at Costessey – The Mill House* – must have been a happy one for Curtis for he had access to a fine collection of entomological books, a well-named (for that day) assemblage of insects and a sympathetic friend and master. The two of them founded an entomological society of which Wilkin was president and Curtis secretary. Among about twenty members were some important people such as the scientific clergy: the Reverends W. Kirby, J. Burrell, Joseph Sparshall and Messrs Brightwell, Joseph Hooker and John Lindley.

Wilkin, who had been studying Latrielle's *Genera Crustaceorum*[44], was much fascinated by that naturalist's classification system and inspired his protégé with an equal admiration for the Frenchman. Curtis now determined to describe, and draw, all the genera of insects, a tremendous task. He lost no opportunity of dissecting and drawing the kinds he could obtain and of copying pictures of foreign insects from the famous continental books. To this end he decided to 'learn a trade' and become an engraver like his father; consequently he got a place with a

* This house no longer exists. A Victorian mansion (1854) has replaced it.

Mr Edwards, an engraver living at Bungay on the Norfolk–Suffolk border. His life was thus divided between two centres, Costessey and Bungay, with the city of Norwich between them; a distance of about twelve miles separated these two places. One wonders how he made the journey: presumably coaches and market waggons traversed these roads or Wilkin provided a horse for him to ride; in any case a twelve-mile walk was not considered excessive in those days, but, walking or riding, one can hardly believe he did it there and back every day.

Wilkin was a printer in Norwich and his patronage of the young man may not have been entirely disinterested. Insects were becoming fashionable among both the public and biologists and this could well mean more books, and illustrated ones at that, so to have a fine artist and good engraver in one's pocket was not a bad thing from a business point of view, in addition to which they were firm friends with the same burning interest. One wonders why Curtis ever left. Perhaps Curtis wanted to marry (Wilkin did not marry until 1825) or perhaps the idea of London with its famous men lured Curtis away.

Wilkin certainly introduced Curtis to many of these famous, perhaps one should say relatively famous, people for they were only known among naturalists. One of these was the aforementioned Reverend William Kirby who must have been impressed by the young man for he gave him the job of drawing and engraving the plates for his joint work with the Reverend Wm. Spence,[41] *An Introduction to Entomology*, the book that ran to seven editions by 1878 and did so much to popularize entomology. On the other hand, though Curtis's plates were very good, Kirby was offered them for nothing by Wilkin, who is thanked in the preface of the *Introduction*. Kirby and Spence had no idea their book would be so successful and would have jumped at the idea of getting the plates free as they are a most expensive item, particularly when hand-coloured, as many of them are. In the introduction the following is found:

In acknowledging their obligations to their friends, the first place is due to Simon Wilkin, Esq. of Costessey near Norwich, to whose liberality they are indebted for the numerous plates which illustrate and adorn the work; the whole of which have been drawn and engraved by his artist Mr John Curtis, whose intimate acquaintance with the subject has enabled him to give to the figures an accuracy which they could not have received from one less conversant with the science. Nor is the reader less under obligation to Mr Wilkin's liberality than the authors, who, if the drawings and etc. had been to be paid for, must necessarily have contented themselves with giving a much smaller number.

2 *Patronage*

In the seventeenth and eighteenth centuries science and art, as has already been remarked, were mainly carried on under patronage. The aristocrat or the rich man had a staff ranging from household servants at one end to his bailiffs at the other, and amongst this body of servants would be found certain professionals, his barber and surgeon, his doctor and chaplain or confessor, his children's tutors, and quite possibly an artist, author, poet or scientist too, if the man's tastes inclined that way, to say nothing of his guards, strong-arm men and so forth, if we may believe some of the sixteenth and seventeenth century plays. ('Enter First and Second Murderer' etc.) Shakespeare's plays, for instance, were nearly all performed by companies under the patronage of the great: the Queen's Company, the Earl of Leicester's Men and so forth.

The people patronized usually resented it, and we have already quoted Dr Johnson in this respect, for they were intelligent and skilled, knew it and feared that these qualities were at the mercy of the whims of a great man, as mostly they were. But the industrial revolution slowly broke up this system. The scientist applied his knowledge and became the technician; whereas Francis Bacon had sought to 'discover the secrets of nature' simply because he was fascinated by the whole working of the world, James Watt sought to apply his knowledge of heat, steam and gases to practical ends, such as the invention of a non-condensing steam engine which, from its greater efficiency, led to such economies in the pumping of water from mines that he needed no patron. Like the man with the better mousetrap, the world sought him out and in this particular case he and a keen businessman (Boulton) made a fortune from it.

What had once been a craft – a 'mystery' in both senses of the word, for the craftsman jealously guarded the secrets of his craft – started to become applied science, that is technology.

Moreover, the general attitude of the aristocratic world began to change. The aristocrat in the fifteenth and sixteenth centuries rather despised the scientist, artist and author. War, hunting, power-seeking and intrigue, with a little religious observance thrown in, were the occupations of a gentleman and he employed 'philosophers' and engineers to fill in idle moments, and possibly come up with a useful service from time to time. But two things started to bring this state of affairs to an end – boredom and the use of gunpowder. It became obvious to the intelligent, more numerous than one might think, that nobody won wars, that intrigue, especially sexual intrigue, was boring and risky. Shakespeare voices the following view of war:

Hamlet enquires of a captain what the army passing by is for:

Hamlet: Goes it against the main of Poland, Sir, or for some frontier?
Captain: Truly to speak, and with no addition,
We go to gain a little patch of ground
That hath in it no profit but the name.
To pay five ducats, five, I would not farm it. . . .
Hamlet: Why then the Polack never will defend it.
Captain: Yes, it is already garrison'd.[69]

To a number of people the study of the natural world offered a more interesting alternative, which our second factor reinforced. Gunpowder meant that war was no longer simply a dangerous sport for king, knight and captain but a conflict in which you were just as likely to be killed as any common soldier. The musket bullet penetrated armour; whereas previously the knight, unhorsed in a mêlée was at once recognized by his armour, shield and standard, and being worth much ransom money, was captured to this end and not killed; now, after gunpowder, war was but a scant respecter of persons.

It took a long time for belief in the scientific theories expressed in the Greek and Latin classics to die, but a series of debunkers, starting with Francis Bacon, gradually discredited these ideas; by Curtis's day science was making great advances. This is not to say that war, power struggle and intrigue had been abolished – Curtis grew up with Napoleon battering around Europe – but that the attitude towards the technician in all its aspects was changing.

The painting of portraits for family satisfaction and of landscapes and other pictures to decorate rooms and for aesthetic pleasure was an expensive process. It takes a long time to paint a picture, even to copy one, and, however poorly paid, 'one must live' as Voltaire had the man say to the Comte d'Argenson. Hence artists tended to work under the protection of a patron, or by appealing to the rich. Romney, for instance, as a bright young man at Kendal, would charge two guineas for a head – a considerable sum – which rose to four in 1763, when he came to London, and was soon thirty-five guineas as he became fashionable – an enormous price for those days (could one say some £1,400 of today's money?). Figures of this nature were beyond the means of the rising middleclass, who nevertheless wanted works of artistic merit in their homes. The engraver began to supply this need, and at the time of Curtis's birth there was a considerable body of engravers at work to supply this market. Gillray's coloured and scandalous caricatures were now sold for a few shillings each.

The engineers were the first to break down the patronage system, for the advantages of their inventions were so very obvious – new roads, bridges, steam engines and firearms. The biologists followed more slowly and here John Curtis can be considered one of the early breakers, in that he used his knowledge and skill in his chosen field – entomology – to break away from patronage and to earn his own living. He was fortunate in a way because he managed to make his intellectual interests provide his bread and butter, but, like most pioneers, he had a hard time. It was but bread and butter he got; very little of life's cake came his way.

There were three stages in his break-away. At first, as we have seen, he was a young man under the patronage of the wealthy Simon Wilkin. In effect, it does not appear that as Wilkin's tame scientist, Curtis suffered slights or insult as Wilkin was an agreeable, kindly person who furthered the young man's career by introducing him to important scientists of the age.

The second stage was when Curtis came to London as a professional entomologist. His main source of income was in satisfying the collecting mania of the age. He also carried out dissecting, drawing, engraving, taking impressions and colouring them for various scientists, such as W. Swainson of Liverpool, to whom we shall refer again. Every gentleman with any pretensions to biological knowledge had to have his 'cabinet' which he would fill with objects in his particular field, such as shells, birds' eggs, insects, fossils and so on. The provision of the cabinets themselves and certain specimens with which to fill them gave

Curtis an opportunity to live from his particular bent. He also wrote and lectured on his subject, though it is doubtful if this last brought much income. But Curtis was more than just an entomologist; being intelligent, poor and having a family he had to be an economist and this led him to the third stage of his career – the describing in a 'practical and useful way' of insects and the damage they do to crops. In other words he became a pest-control technician, but one with a remarkably sound knowledge of the basic principles and details of his craft. The first half of the nineteenth century saw considerable improvements in agriculture: enclosure, a better knowledge of how plants lived, the Norfolk four course rotation, and, above all, better manuring, were important factors in improving yields. However mistaken the 'humus theory'* of plant growth was, and it was much believed in by practical men, it did have the effect of leading farmers to conserve their manure, to tread straw and to use the liquids running from the middens, thus adding additional quantities of the much needed nitrogen, phosphate and potash, the main foods plants obtain from the soil. In the late 1820s guano was being imported from Peru, and Chilean nitrate of soda soon followed. All this led to increased yields which in their turn inevitably lead to greater losses from pests. Even if the percentage of loss from pests does not rise, the larger yield makes the damage more noticeable. For instance, in 1800 a 15 per cent loss due to, say, wire worm, bunt, take-all and wheat bulb fly on a 2-quarters per acre crop of wheat is 150 pounds, 2½ bushels, regrettable but usual; on the other hand a 15 per cent loss on the 4-quarter crop of 1840 is 300 pounds and is much more obvious. Moreover the percentage loss is likely to be considerably higher than this if conscious or unconscious steps are not taken against the pests. A bigger supply of food, say wheat, all concentrated in one spot, is ideal for those forms of life that live on wheat, and consequently they (including man) tend to accumulate there in order to garner the crop so that losses start to rise both in actual and in percentage terms. Of course yields and losses from pests vary enormously from year to year and the exact amount of damage done by pests was but little recognized in Curtis's early

* A theory promoted chiefly by Albrecht Daniel Thaer and Mathieu de Dombasle that plants obtained their carbon from decomposing plant and animal organic matter in the soil. It was quite wrong: the carbon is in fact obtained from the carbon dioxide gas in the air.

days. In fact, in economic terms it is but little realized today and we shall have more to say about this in chapter 8 and appendix 6. Curtis in his third period was well aware of these losses. In his introduction to *Farm Insects*[17] he writes:

> Whilst Chemistry and Geology have lent their valuable assistance in rendering the land more fertile, little attention has, comparatively, been paid to those noxious animals which annually consume an amount of produce that sets calculation at defiance; and, indeed, if an approximation could be made to the quantity thus destroyed, the world would remain sceptical of the result obtained, considering it too marvellous to be received as truth.

3 *Entomology in the early Nineteenth Century*

Seventeenth and early eighteenth century biology had been much bedevilled by the difficulties inherent in the naming of plants and animals. There was no standardized system; each worker gave a name of his own invention to the creature or plant he was studying and, whilst correspondence was facilitated by the use of a *lingua franca* (namely Latin), two or more correspondents could never be quite sure they were talking about the same thing. Thus one of the dung beetles was called *Pilularius* by Mouffet[51] (d. *circa* 1604), and he had to give the common names in a number of languages in order to identify the insect. Fabricius (1668–1736) called it *Copris* and Latrielle (1762–1833) *Onthophagus*. The first move towards ending this confusion was made by the famous Karl Linne (Linnaeus is the latinized version, (1707–78)) who invented the binomial system giving one name to the genus and another to the species, a system which persists to this day. Linnaeus called the above-mentioned insect *Scarabeus nuchicornis*. As it happens that Curtis was brought up on Latrielle, he (Curtis) stuck to the latter's name.

The naming of new species of insects was one of the passions of the early nineteenth century collectors; one might even get a friend to name a creature after oneself and thus hope to achieve immortal fame. For instance, *Bledius Skrimshirii* was named by Curtis in honour of 'its captor, the Reverend T. Skrimshire, whose zeal and liberality have enabled us to present our readers with this striking novelty'.[13] It is a beetle which lives on the sea shore, particularly in East Anglia. We know but little more of the learned, zealous and liberal cleric. Alas, in many cases these names were evanescent; they were replaced by others under the priority rules, or because they have been put in the wrong genus, and even this little fame is denied them. For instance, a small ermine moth, named by Curtis as *Spilosoma Walkerii* in honour of Sir Patrick Walker of Edinburgh, first sank to being considered a mere variety of *S. menthastri*, the white ermine moth, and subsequently became *S. lupricepeda*, two removes from Curtis's

Name used by Curtis	*Vol.**	*Plate No.*	*Date*	*Commemorating*	*Present Name*
Coleoptera *Helobia Gyllenhali* Leach	I	103	1.2.26	?	*Nebria gyllenhali* (Schönh)
Leiochiton Readii Curt.	I	346	1.3.31	Rev. J. B. Reade	*Miscodera artica* (Payk.)
Pogonus Burelli Haworth	I	47	1.12.24	Rev. J. Burrell	*Pogunus luridipennis* (Germ.)
Hydroporus Davisii Curt.	I	343	1.12.31	A. H. Davis	*Oreidytes borealis* (Gyll.)
Eniccerus Gibsoni Smith	I	291	1.1.30	Mr Gibson of Halifax	*Ochthebius exsculptus* (Germ.)
Dendrophilus Sheppardi Kirby	I	131	1.9.36	Rev. Revett Sheppard	*Dendrophilus pygmaeus* (L.)
Lesteva Leachii Kirby	I	303	1.4.30	Dr Leach	*Deleaster dichrous* (Grav.) †
Bledius Skrimshirii	I	143	1.12.26	G. Skrimshire & the Rev. T. Skrimshire	*Bledius furcatus* (01)†
Callicerus Spencii Kirby	I	443	1.3.33	Rev. W. Spence	*Calicerus obscurus* (Grav.)
Lepidoptera *Arcturus Sparshalli* Curt.	V	336	1.12.30	Joseph Sparshall	*Trichiocercus sparshalli* (Curtis)
Spilosoma Walkerii	V	92	1.11.25	Sir Patrick Walker	*Spilosoma luteun* (Hufnagel)
Apamea Haworthii Nob.	V	260	1.5.29	Mr Haworth of Edinburgh	*Celaena haworthi* (Curtis)
Melanipe Blomeri	VI	416	1.8.32	Capt. Blomer	*Discoloxia blomeri* (Curtis)

* The volume numbers quoted above are those in the systematic arrangement of John Curtis's *British Entomology*.

† Still retained as a varietal name.

friend and customer. On p. 15 is given a list of some of these old names and the newborms that have replaced them.

Nevertheless entomology was popular among 'natural philosophers' in the mid-eighteenth century, the interest in it being divided between scientifically minded biologists and trophy hunters. An example of eighteenth-century interest is the famous sale of the collection of that patroness of Natural History, the Duchess of Portland. It started on 24 April 1786 'and the thirty-seven following days' as the catalogue puts it! Of course, not all the lots were insects, but many were: Dale tells us[22] there were some fine Dorset *Papiliones* in the collection. One cannot but admire the endurance of all concerned, buyers, auctioneer and staff and hope they got Sundays off. The rising middleclass had not the money to go on the grand tour, and bring back pictures, statues or rare game animals, alive or dead, but some of its members could discover unusual insects and some might even have the creatures named after them. *Acherontia Atropos* could be a substitute for a lion skin and a moth bearing a death's head on its back and able to squeak on capture was a creature to stir the imagination; one could weave legends around it.

The first entomological society in the world was formed in London in 1745. It was called the 'Aurelian Society' and one Moses Harris was active in it. It met at the Swan Tavern, Change Alley; but little is known of it for all its records were burnt in the Cornhill fire of 1748. From its name one assumes it was more interested in butterflies and moths than in other insects, though Moses Harris's book has more about Diptera than any other order. The name 'Aurelian' may also have been used with the idea of connecting the society with Marcus Aurelius Antoninus, the philsosopher-emperor (AD 161–80). After the fire the body collapsed, but a second Aurelian Society was formed in 1762 and in its turn faded away; and a third started in 1801. From 1780–2 there was an Entomological Society of London and desultory meetings were carried on until 1822. In 1826 the Entomological Club was formed and the present entomological society (now the Royal Entomological Society) was instituted in 1833.[54]

In 1810 there was a Norwich Entomological Society in which John Curtis was active, and to which we have already referred; in fact we may be pretty sure that it was this body and the meetings with Simon Wilkin there that drew Curtis away from the law and into his career. The third meeting of the Norwich Society, 3 December 1810, was held in 'the house of Mr John Curtis', presumably his mother's house – the 'cultivator of flowers' in the Cathedral Close. Curtis at this time would have been nineteen years old. It is, of course, possible that the meeting was held in Curtis's rooms in Simon Wilkin's house. We can thus see that entomology was much in the air and Curtis may be said to have taken to entomology as a profession when he went to live with Simon Wilkin, although at that time he may still have had some hope of becoming an artist, for he exhibited pictures at the Norwich Society of Artists up to his departure for London, as follows:

1810	No. 108	Moth, from nature
	185	Butterfly, from nature
1813	144	Flowers
1817 Spring	116	Original drawing for the *Introduction to Entomology*
	180	Twenty-four rare British Insects, a coloured engraving
Autumn	25	Head of Christ, after Titian
	34	Pair of birds—the bearded titmouse

Curtis continued to exhibit at the Society's shows from London for a few years. From his choice of subjects one could assume that his heart was still in the beautiful Norfolk countryside. The Society's catalogues have the following entries:

1818	No. 111	View near Bungay, Suffolk
1819	104	View at Hetherset, Norfolk
1820	151	Flower piece, Chrysanthemums

Unfortunately none of these exhibits appear to have survived.

Curtis's transfer to Costessey took place about 1813 and at that time, apart from the classical works, such as those of Mouffet,[51] Reaumur,[65] Latrielle,[44] Buffon,[98] Cuvier[21] and Linnaeus,[47] which were so expensive that only the wealthy could purchase them, there were very few popular books on entomology. Even Millar's *Natural History* of 1785[50] cost new £2 8s., say £50 of our money, with the author maintaining that it was a cheap publication.

The great change that occurred in the mid-nineteenth century was cheap printing, which enabled the rising middle classes and the working man to instruct themselves should they so wish. Books, newspapers and periodicals became things almost everyone could have. Literacy among the population greatly increased and this tendency was perceptible as early as 1820. The eighteenth-century public for the printed word was much more restricted.

Certainly from 1818 onwards, perhaps before this too, to some extent, the young Curtis, through Simon Wilkin and other friends, would have had access to the books of the great masters mentioned above, works usually of several volumes, which were bound in calf or tooled leather and stamped with one's arms, if one had any or could invent them. Slightly cheaper and more ordinary works were becoming available, however, and for some reason a number of the moralizers of the age seemed to think that botany and entomology were particularly suitable for training the young.[66] 'A Lady' in 1797 published *Botanical Dialogues between Hortensia and Her Four Children.*[42] In 1816 Priscilla Wakefield was instructing youth with *An Introduction to the Natural History and Classification of Insects*,[82] in what purports to be a series of letters from Felicia to Constance. Alas! we do not have any of Constance's reactions or replies and cannot know whether Felicia's contention that, for the young, 'Nonsense has given way to reason: and useful knowledge, under an agreeable form, has usurped the place of the Histories of Tom Thumb, and Woglog the Giant' was true or not, but one feels for the children struggling with eight kinds of antennae (*barbutæ* to *setaceæ*) and similar technicalities rather than following the wonderful stories of giants and dwarfs. Priscilla drew on George Shaw[70] and Barbut[2] for her material, as she acknowledges, and one wonders if the work was not really by Shaw himself, in an attempt to supplement his income.

The tendency was still continuing in 1825; an anonymous author produced *Letters on Entomology intended for the Amusement and Instruction of Young Persons and to Facilitate their Acquiring a Knowledge of the Natural History of Insects*, a mouthful, one would have thought, sufficient to put a child off the subject for life. This publication is also supposed to be a series of letters from a lady to her sister Harriet; again most of the observations seem to have been taken from George Shaw's large *General Zoology*, but the author has not plagiarized Priscilla Wakefield.

Nevertheless, even by 1816, a few technical books on entomology had been produced at medium prices, the forerunners of the flood of cheap works to be released on the Victorian age. The most important was Kirby and Spence's *Introduction to Entomology* (1815),[41] about which we shall have more to say later, but as it was eventually in five volumes it can hardly be called cheap, although it became immensely popular. Earlier works were the Reverend W. Bingley's *Animal Biography* (2nd edn 1814), in three volumes and that again was not cheap; a third of the third volume was on insects.

George Samouelle published *The Entomologist's Useful Compendium*[68] in 1819, and particularly mentions that his work is for those who cannot afford the four volumes of Kirby and Spence, 'yet its extent, and the four octavo volumes, must exclude many from purchasing it, and especially young persons to whom the study of Entomology is particularly adapted'. Samouelle's work has twelve very good and clear plates; there is no indication as to the artist or engraver and one suspects, from their style, that both (or at least the engraver), may have been John Curtis. Kirby and Spence's *Introduction to Entomology*, published in 1815, was the book above all others which gave most impulse to the study of entomology. Kirby and Spence thought that the public regarded them, and entomologists in general, as ridiculous, old gentlemen rushing around with butterfly nets and magnifying glasses in pursuit of trivialities. In fact this was probably not the public's opinion at all, but only what the naturalists imagined it to be, a view confirmed from my own experience many years ago in Argentina. A colleague and I were studying locusts in a remote part of the Pampa during a tremendous invasion of that part and were endeavouring to find out if certain mixtures were attractive or repellant to the crawling millions. We would put a line of numbered plates out, each with different bait mixtures on it, in front of an advancing swarm, and every two or three minutes, for days on end, we would count the number of locusts on each plate. But if one got near enough to a plate to count the locusts they took fright and jumped off, and so we used field glasses for the purpose. People came to look at us and we thought that if watching a dozen or so plates through field glasses for a week did not qualify us for admission to the nearest mental asylum, or immediate expulsion from the country, nothing would. But the contrary was the case; the local villagers regarded us as great scientists and believed that we would abolish the plague for ever – a situation even more embarrassing than the first.

To return to Kirby and Spence's *Entomology*, these gentlemen's fears were expressed as follows:

One principal cause of the little attention paid to entomology in this country, has doubtless been the ridicule so often thrown upon the science. The botanist, sheltered now by the sanction of fashion, as formerly by the prescriptive union of his study with medicine, may dedicate his hours to mosses and lichens without reproach; but in the minds of most men, the learned as well as the vulgar, the idea of the trifling nature of his pursuit is so strongly associated with that of the diminutive size of its objects, that an *entomologist* is synonymous with everything futile and childish. Now, when so many roads to fame and distinction are open, when a man has merely to avow himself a botanist, a mineralogist, or a chemist – a student of classical literature or of political economy – to ensure attention and respect, there are evidently no great attractions to lead him to a science which in nine companies out of ten with which he may associate promises to signalize him only as an object of pity or contempt.

If this attitude were ever at all widespread, Kirby and Spence, and other entomologists, Curtis among them, certainly changed it. The *Introduction to Entomology* became a best-seller; by 1856 it had run to six editions, with a seventh, the *People's Edition* (*Fifteenth Thousand*) appearing in 1878, some time after both the authors were dead. Collectors multiplied and books were published as never before. We may note, in passing, that a caricature of the kind feared, and possibly conjured up by Kirby, can be seen in the frontispiece of an anonymous book, *The History of Insects*, London, 1842 (The Religious Tract Society).[67] It contains a rather poor, hand-coloured plate showing an elderly gentleman, pale blue trousers, black tail-coat and top hat, chasing a butterfly with a two-handed net whilst a boy, in a green short-skirted garment and gaiters, dips a net into a pool. Lying on the ground are various bits of entomological apparatus and another top hat, presumably the boy's. One wonders if perhaps it is not a plate borrowed from some French or German publication.

The early examples of Curtis's work show great talent combined with practical utility. For instance, by the side of the picture he places a line giving the true size of the insect. In Kirby's *Introduction* Plate III is the most attractive, showing a butterfly and two moths, two dragonflies and a stonefly. One of the moths, the crimson speckled footman, is particularly well done, but then the creature lends itself to a striking plate having a pattern of red and black dots on a cream coloured ground on the fore wings and white rear wings with brownish-black edging. It is a migrant to Britain, the caterpillars feeding on forget-me-not and borage. Among its scientific names are *Bombyx Pulchella*, Kirby, *Deiopeia pulchella*, Curtis, and *Utetheisa p.*, Hübn and Linn.

Of great concern to entomologists of that time (around 1817) was the classification of insects. The majority of them were firm believers in the special creation of each species as laid down in 'The Book of Genesis', but this did not prevent them seeing relationships between species and genera – a heresy that was eventually to lead to Darwinism. That there *is* a relationship is fairly obvious and indeed this had been noted even in classical times. 'Aristotle had in mind some systematic arrangement which through various stages advanced to a definite system in the work of Aldrovandus and eventually led to the Linnean system of binomial nomenclature.' (Wilson and Dorer).[89] John Ray (1628–1705) established the species concept, which helped forward the matter and Buffon (1707–88), agreeing on the whole with him, was unwilling to recognize any systematic unit higher than the species. In Curtis's early years the main rival systems were those of Linnaeus, Cuvier, Latrielle and Fabricius. Linnaeus had the right basic idea (the binomial system) but tried to classify insects according to their wing structures, which was not very practical. Cuvier and Fabricius based their systems on the comparative morphology of all organs, but principally on those of the mouth. Latrielle described his system as 'natural' and 'eclectic', that is a system which combined features from all the other systems. Curtis, early in life, came under the influence of Latrielle, and so always used this system and Latrielle's nomenclature. Modern insect classification is based on Handlirsch (1908) as modified by Brues and Milank (1932); it descends from Latrielle's natural system but with thirty-four orders instead of Latrielle's final modification showing twelve.

4 *Curtis Comes to London*

We might now pause to enquire what the world was like when Curtis came to London in 1817 or 1819. It was much dominated by Waterloo and the fall of Napoleon, which the British preferred to dwell on rather than their own defeat by the USA at New Orleans. George III was still alive, but insane, and the 'First Gentleman in Europe' was regent. Louis XVIII had been restored to the French throne, and had shot some of Napoleon's more irritating marshalls (Murat and Ney, for instance) obviously taking Voltaire to heart,* for nature perpetually imitates art; Pope Pius VIII had restored the Inquisition, so that all seemed set for a brilliant future for the ruling classes as the industrial revolution gathered force. In 1814 George Stephenson had produced the first effective locomotive and the first railway accident shortly afterwards led to the railway age which expanded the market for manufacturers. In 1818 the steamship *Savannah* crossed the Atlantic in twenty-six days. In 1819 the battle of Peterloo took place: a large crowd of workers demanding reform were dispersed by cavalry, five persons were killed and many wounded. As a result the Six Acts were passed against 'seditious meetings'; nevertheless certain reforms were made, such as the act laying down a maximum twelve-hour day for child workers (including Saturdays), one guaranteed, so it was said, to ruin the nation.

Important literary works were coming forward. Byron 'awoke one morning and found himself famous' (*Childe Harold*, 1812, 1816, 1818). The Waverley novels were started and women, moving slowly towards emancipation, showed they could be as good (or better) novelists as men; though some found it prudent to disguise themselves with male or neuter names (George Eliot, Currer Bell). Jane Austen was successful simply writing as a woman. Serious economic studies were made, such as William Cobbett's *Paper against gold: the history and mystery of the Bank of*

* 'Dans ce pays ci il est bon de tuer de temps en temps un amiral pour encourager les autres.' Voltaire, *Candide*, chapter 23.

England (1817) and Ricardo's *Principles of political economy and taxation* in the same year. It is interesting to note that William Spence, joint author with William Kirby of the *Introduction to Entomology*, was also an economist and produced a number of *Tracts on Political Economy*,[73] many running to several editions.

The corn laws were the great political controversy of the age. They had the effect of keeping the price of corn high for the benefit of farmers and land owners. During the Napoleonic wars wheat was seldom less than £5 per quarter (£22. 8s. per long ton), about the same price as it is today in our much depreciated money.

When Curtis came to London the industrial age was getting into its stride. Many historians regard it as one of increasing prosperity and development, and so it was, but only for about 15 per cent of the population. (History has but little to say on how majorities live).[61] Increasing numbers of working people were crowding into the cities to man the new industries; slums grew and the sweat-shops started. Conditions for the poor got steadily worse. Henry Mayhew studied the matter and we report here from an article of his of 1850 on three young girls doing 'needle and slop work'.

No. 1 made moleskin trousers. 8d. per pair. When fully employed earned 8s. per week. She paid 1s. 6d. for thread and candles, coals 6d., rent 1s., balance 5s. Average net balance 3s. per week on which to feed and clothe herself.

No. 2 made slop trousers; worked till nine or ten every night. Cleared 3s. per week.

No. 3 made 'fine full-fronted white shirts' 2¼d. each, six buttonholes and buttons, four rows of front stitching, fix collar and waist bands. Working from 5 a.m. till midnight did seven per week.

Of course the money was worth more in those days, but even so the wages were appalling and the hours of work speak for themselves. It is not surprising that many gave up this slavery in favour of prostitution. It might lead straight to hell, which could not really be very different! *Laissez-faire* economics, however, gave the comfortable assurance that any interference with the system would merely make matters worse.

The succession to the throne was another problem hovering in the background even at this early date. The regent's only child, Charlotte, died in 1817 and, separated from Queen

Caroline, he was unlikely to have any more. His four unmarried brothers were rapidly married, but only the Duke and Duchess of Kent had a legitimate child, Princess Victoria, born in 1819, who was destined eventually to give her name to an age.

Agriculture and agricultural science were still pretty backward. In the former, yields had been increased by an empiricism, such as better rotations, which science was later to explain and improve. Very little was known of how plants fed (see p. 12). Between 1803 and 1813 the Board of Agriculture had earned the gratitude of many farmers by organizing a series of lectures by Sir Humphrey Davy on plant nutrition; these talks formed the basis of his book *Elements of Agricultural Chemistry*, 1813,[24] showing the importance to plants of the surprisingly inert element nitrogen. The impetus given to scientific farming was but short-lived, for the government abolished the Board in 1820 as an economy measure, and interest and work on the subject did not revive until the formation of the Royal Agricultural Society in 1838. Under the patronage of the young queen: then, and ever since, it has lived up to its motto 'Science with practice'.

It remained for Leibig (1840), guided by Ingenhousz (1801), really to uncover the plant cycle. Leibig had a great belief in careful analysis: over the door of the university he founded he had carved 'God has ordered all His Creation by Weight and Measure', and using this principle and his belief in the indestructibility of matter he uncovered the true facts of plant nutrition: that the carbon came from the air and the minerals from the soil.

Such then, roughly, was the state of biology applied to agriculture when Curtis set out for London; but to this have been added a few notes on its subsequent advance. Curtis's establishment in London is variously put at 1817 and 1819. In any case he made the journey via Barham, near Ipswich, the residence of the Reverend William Kirby. Possibly he stayed a considerable time at the rectory, with occasional visits to London, and only settled in the capital in 1819. It may well even be that Kirby 'stole' him from Wilkin in order to have a skilled artist on hand to illustrate his books and arrange his collection of insects. Alternatively there may have been considerable cooperation between Costessey and Barham, with visits to the engraver Edwards at Bungay, because Kirby in his introduction to the

Introduction acknowledges the help received from both Wilkin and Curtis.

At Barham Curtis met Kirby's co-author Spence and also William S. MacLeay, who later became of great use to the young man. In 1814 or 1815, whether at Costessey, Barham or both, Curtis helped Kirby dissect insects and made the drawings, or some of them, and the first three plates (hand-coloured) for the famous *Introduction to Entomology*. The next two (Plates IV and V) were his and are dated 1 January 1817 and are also coloured. The following nineteen plates are certainly in his clear bold style. They are mostly of dissected parts of insects and appeared much later on. The final six plates (XXV to XXX) are by another hand – H. Denny (1803–71). It is curious that the change should have been made and we will come back to this later, merely noting here that the Denny plates are much in Curtis's style. Perhaps he left some sketches for his successor, if not the actual drawings.

The etcher needs quite a bit of heavy apparatus – copper plates, tools, baths and bottles for acid and above all a press, for he must take samples of his work, so one wonders where Curtis made his prints. He could hardly lug all this stuff between Costessey and Barham, some fifty miles, and to London too, and he may well have prepared the plates at Bungay where he learnt the art from Mr Edwards. Indeed the need for a settled workshop and studio of his own could have been an inducement for him to go to London, as well as the excitement of being at the centre of scientific thought and meeting the famous in this field.

Introduced by his new friend MacLeay to the Linnean Society, Curtis was soon doing plates for them. In Volume XII of their *Transactions* (1817–19) we find that from Plate IX onwards Curtis has replaced Weddell as their main engraver and that the volume is then enriched with much better illustrations. Curtis even engraved some plates (Nos. IV–VIII) from drawings by Weddell, which suggests there was no great rivalry between these two men. The 1819 date makes it look as if Curtis came to London at this time. Weddell returned in Vol. XV and did a few plates.

This same volume also contains Kirby's famous *Century of Insects*[40] for which Curtis did three magnificent plates, showing forty-five insects, most of them Coleoptera, many collected by Robert Brown in Australia.[6] Kirby describes the insects as 'nondescript', using the word in its literal sense – the insects had not been described. He seemed to think that British entomology was in a poor way and this may have moved him to help promising young men, such as Curtis and Denny. In the *Century of Insects* he writes:

The infinite hosts of nondescript insects occurring in the entomological cabinets of this country, is rather disgraceful to us: and what is more so, we permit foreigners to do that for us which we are competent to do ourselves. Witness the numbers described by Fabricius, Olivier and others from British collections.

Pudet haec opprobia nobis
Et dici potuisse et non potuisse refelli

I am not, however, so illiberal as to wish that the entomologists of the continent should be excluded from our cabinets: if we ourselves are too idle, or too busy, to give the public some account of our entomological treasures, it is for the interest of science that they should do it for us.

Once in London, certainly by 1819, Curtis took lodgings from Miss Dorothy Western, 12 Charles Street, Middlesex Hospital. It can be thought of as a reasonably select area as Miss Western was paying a considerable rent – £100 per annum and rates of £13. 15s. A modern historian, Ann Cox-Johnson[12] has a fairly large number of people in No. 12, suggesting that Miss Western was packing her lodgers in to pay the high rent – Messrs F. Cruickshank, J. W. Faulkener, R. Ramsey, P. Reinagle and J. H. Savage and the Misses Rosa Drummond and Mary Webb, some of the above, no doubt, being servants. But Miss Cox-Johnson has John Curtis and James Minasi as living at No. 13 Charles Street, which is incorrect as far as Curtis is concerned: there may have been a switch of numbers at some point or Curtis may have had a studio and engraving shop over the road.

Kirby had reason to be grateful to his young protégé and sought to forward his career; he introduced him to Sir Joseph Banks (1743–1820), President of the Royal Society. Sir Joseph allowed the young man the run of his library and invited him to his learned *soirées* where all the principal scientists met. Here he met Dr Leach (1790–1896), superintendent of the Zoological Collection of the British Museum, and in his house Curtis met

J. C. Dale who later did much to help the rising man. The similarity of Curtis's and Leach's ages and tastes, both of them being admirers of 'Citizen Latrielle' (Appendix 2), led them to become firm friends. Leach also presented Curtis to Baron Cuvier, Bonelli and the great Latrielle himself when these gentlemen visited England. Their joint admiration of the latter led the younger man back to his project of describing all the British Insects and thus perfecting the classification system. Curtis began to dream of his great work, *British Insects*. Leach had also inspired Curtis with an interest in shells and their structure and an expedition to Scotland in order to drag 'the whole coast' for molluscs was proposed. It seems a colossal undertaking; one can hardly suppose they did not realize what an enormously long coastline Scotland has. Perhaps Leach's mind was already beginning to give way and brush over inconvenient details, and Curtis was only too glad to be associated with an interesting and probably, for him, profitable undertaking. At any rate, Curtis, now twenty-nine or thirty, must have felt that things were turning out well for him and it is likely that he married for the first time in 1819 or 1820. Very little is known about this marriage: a son Edward John was born at Paddington on 20 March 1821 and that is about the sum total of our knowledge on this subject. The gaps in the Curtis story are many. The first Mrs John Curtis is such a shadowy figure that one wonders whether she may have died at the birth of their first child, but, though plausible, this is unlikely for there are two other clues to her – the 1841 census and the Bank of England Archives. In 1841 the Curtis household consisted of himself, Lydia Curtis, aged forty, Edward and Elizabeth Curtis aged respectively twenty and fifteen, and Anne Turner, aged seventeen. One presumes Lydia was his wife, Edward and Elizabeth their two children and that Anne was the maid. The Bank of England archives show that Edward John Curtis when twenty-one years old (in 1842), obtained a post with the Bank of England and it is recorded that he had 'quitted school four years, was three years studying at home with his father with whom he resides in the Hampstead Road, and will continue to do so'.

At the time of the birth of his son, Curtis was probably living at 4, Grove Place, Lisson Grove, London.[64] He was certainly there in 1831, when he was paying a rent of £50 per annum and rates of £10. 4s. 2d. The make-up of the rate is of interest:

	£	s.	d.
Poor Rate	6	13	4
Repairs, lighting and cleaning streets	2	3	9
New paving		6	3
Metropolitan Roads Act		4	2
New churches		16	8
	£10	4	2

One wonders if Jews, Dissenters and Roman Catholics were obliged to contribute to the new Anglican churches. If so it must have been a charge that rankled. The poor were obviously a burden at this time; 65 per cent of the rates were devoted to their support, a far higher proportion than is given to 'social security' today. Curtis seems to have been a good payer, never in arrears and in fact he overpaid on the above occasion by £1. 12s. 8d., in two instalments of £5. 18s. 5d., which seems strange as he was always hard up.

In the event things turned out badly for the new Londoner. Leach's mind gave way and he retired from his post in 1821 and went to live with his sister at Tortona, Italy. The Scottish tour was abandoned. It was a severe blow for Curtis. Friends rallied to his aid, particularly Mr MacLeay, who introduced him to Dr Sim (1799–1831), the botanist and physician who edited William Curtis's *Botanical Magazine* and who gave John Curtis a certain amount of work. The two Curtises, William and John, were in no way related; the former died in 1799 when John was eight years old.

Curtis never gave up being a scientific entomologist, but in order to live he had to make that science supply his livelihood. It seems very probable that about this time he began to sell insects to collectors. In 1822 he was made a fellow of the Linnean Society, a tribute to the high respect in which he was held. He began to make serious plans for *British Insects*, in which he was much aided by the thought that Kirby had prepared the way, for the *Introduction of Entomology* had run to three editions even by 1818. The subject having been introduced, the time was ripe for a serious study.

We have already noted that Curtis took over the plates for Volume XII (1817–19) of the Linnean Society *Transactions* and he continued to engrave for them in Volume XIII (1821–22). Out of thirty plates twenty-six have been done by Curtis and he even drew the figures for five of them, the subjects being various wild dogs and marmots. He was versatile. In this volume two plates are unsigned and two are by J. S. Basire.

Due to his connexion with the famous and with the Linnean Society, Curtis saw himself as a rising scientist, but with no private income or patronage he had to find a way of living and whereas today it is comparatively easy to make commerce pay for scientific research, in those it was unknown. An example of this difficulty is the brush he had with William Swainson (FRS) of Liverpool in 1820. Swainson was a collector and eventually became a founder member of the 1833 Entomological Society. In 1820 he commissioned Curtis to print and colour etchings of shells and insects for him and among the papers at the Linnean Society some of the accounts covering such transactions still exist; we reproduce one here, in some detail, both because of its general interest as to ruling prices, and for its bearing on this matter:

W. Swainson Esq. to S. Curtis

1820	Cr.	£	s.	d.	*1820*	Dr.	£	s.	d.
July 26	By cash	5	0	0	July	Etching tools	1	13	11
Aug. 4	By lot of imprints at Bullocks		17	0	Aug.	Colouring 13 plates, 125 each being 1625 impressions at 17s. per 100	13	16	3
	Balance due to John Curtis	10	5	4		Hotpressing Do 9d per 100		12	2
		£16	2	4			£16	2	4

Curtis was getting 2.13d. (old pence = 0·89 new pence) each for colouring and pressing, which does not seem very much. However, Swainson appeared to be dissatisfied. The reason for this was that Swainson had learnt lithography and, thinking that Leach's encouragement of his proposed book, *Zoological Illustrations*,[79] was also an endorsement of his lithographic powers, wanted to illustrate the work himself in place of Curtis's etchings. Instead of saying so straight out, Swainson seems to have taken a rather underhand way of escaping from his dilemma. He started to complain of Curtis's colouring, of delays and of costs. Probably he was not too sure of his own abilities, and of the time he had available, and thought he might as well use Curtis so long as he was cheap.

Curtis, not knowing of Swainson's artistic ambitions, retorted to these complaints with spirit. On 15 October 1820 he said that one of the reasons for the delay was that Swainson was putting the wrong address on letters to him. He, Curtis, could not reduce the price to 1d.; he was not in this particular business for profit but for friendship's sake and the advancement of science, and if he (Swainson) was not satisfied he should find someone else. Swainson then appeared to have been somewhat cautious, though still aggressive; he did not at that time immediately find 'someone else', that is himself, and seems to have suggested a more extensive commitment with Curtis. On 20 November 1820 Curtis wrote to Swainson saying he was making him work very hard for but little money, and was too severe in his accusations and complaints, that he had now founded a regular establishment to handle Swainson's work and please, please send the material regularly and in time.

This appears to have driven Swainson over the edge and he wrote to say he was now taking the work out of Curtis's hands. To this Curtis replied saying this was most surprising just as he had made arrangements to handle an increased quantity of it; he did not mind for his own account, but for the people he had engaged for it. The sums Swainson paid did not cover his loss of time and the pens, ink and paper. He enclosed another account:

A/c 1820	£	s.	d.
Drawing dissections of insects	2	2	0
Engraving plates, copper and expenses	7	7	0
Printing 130 of do		10	8
Colouring plates 5, 6, 9, 15, 18 and 24 130 of each at £1. 1. 1. per 100	6	6	6
Do Plates 11, 16, 19, 20 and 29 at £1. 3. 2. per 100	5	15	10
Hot pressing at 9d. per 100		10	9
Postage of 5 letters		8	0
	23	0	9
100 of *Colias* at 23s. 2d. per 100	1	3	2
	£24	3	11

The item for postage is enormous, about 1s. 7d. per letter (8 new pence): no doubt we shall soon have achieved this figure again in Britain, but whether we can ever reach the speed of the old Roland Hill days remains to be seen. (Appendix 5.)

Swainson (2 December 1820) disputed the account and tells Curtis to send him all his stuff, as he has made 'other arrangements more satisfactory to myself and more advantageous to the work'. He maintained that Curtis quoted him 17s. per 100 for *Colias* and is charging 23s. 2d. Curtis replied that there was no inconsistency as Swainson had not kept his bargain: he reduced the number wanted from 500 to 250, he sent crowded plates quite unlike the sample submitted* and, as a final fling, Curtis said he had recounted the plates and had amended his bill, increasing it to £28. 1s. 4d.! With commendable caution he also said he would hand everything to Swainson's messenger on receipt of this sum. This defeated the irascible Swainson: he sent a draft on Barclay's Bank for £28. 1s. 4d., saying he trusted Curtis to deliver. On 14 December 1820 Curtis fired the last shot telling Swainson he had sent the parcel to Charing Cross and that this was 'the last time you possibly can be annoyed by your obedient servant John Curtis'.

Needless to say Swainson's lithographs are nothing like as good as Curtis's etchings: in any case etching, with the possibilities of very fine clear lines, is a much better medium than lithography for presenting entomological detail. In fact Swainson went back to etching; in his *The History and Natural Arrangement of Insects* (1840)[79] he uses an etched plate, engraved by one E. Findon, from a drawing by himself as a frontispiece, and woodcuts in the text. He was much concerned with establishing an elaborate classification system for the lower animals, which it would be tedious here to follow and which never received much acceptance among biologists.

* A crowded plate could, of course, be cut up and two or three plates made of it.

5 *'British Entomology'*

Curtis spent most of 1823 getting ready for 1 January 1824, the great day on which the first number of *British Entomology, being Illustrations and Descriptions of the Genera of Insects found in Great Britain and Ireland* was to appear. New Year's Day, 1824, was a Thursday and the five plates making up the number duly appeared. In fact throughout the whole fifteen years this serial publication ran Curtis was never once late, a statement which cannot be made of most scientific journals today. With so much less leisure, so many fewer people and hardly any mechanical devices to go wrong (no dictating machines, no electric typewriters and above all no telephones – the constant source of delay, interruption, uncertainty and confusion) it is no surprise that local business moved more quickly and surely in the mid-nineteenth century.

The preface, which was written in 1839, after the work was completed, sets out the objects of the book very clearly and we quote:

Some years before peace opened to this country the stores of the Continent of Europe, I had studied and admired the classical *Genera crustaceorum et insectorum* of Latrielle, the utility of which was so manifest, that a translation of it was suggested by several talented scientific men connected with the 'Entomological Society of Norwich'; this first led me to consider the best means of supplying a work which might be equally useful to my fellow-labourers at home, and the result was a determination of preparing for the press an *Illustrated Genera of Insects*. After a period of several years, during which I never lost any opportunity of adding to my materials, I found them increase so rapidly, that it appeared necessary to confine myself to the *Genera of British Insects;* with this view I took immediate steps for the publication of my work, which commenced on New Year's Day, 1824.

In other words Curtis had soon discovered that to describe and classify the world's insect genera was too big a task. It should be noted that the original intention was to describe the genera only, one or two descriptions of species being given as examples. But this plan was changed during the course of the work, as we shall describe further on in this book. It must be remembered that this preface was written with hindsight. The somewhat embittered middle-aged man of 1839 was very different from the young enthusiast preparing the first number of *British Entomology* in 1824. Even so the preface does give us interesting views on his reasons for the book. It continues by pointing out that it is strange that such a work had not appeared before, particularly as Gilbert White, in the *Natural History of Selbourne*, fifty years ago, had observed that 'nothing would recommend entomology more, than some neat plates, that should well express the *generic distinctions* of insects according to Linnaeus'.[88]

Curtis then explains his classification system, essentially one modified from that of Latrielle, as has already been noted. He continues by contrasting the assistance the entomologist gets today (1839) with conditions when he started. Then there was no printed systematic catalogue: such an item was so essential that one was compiled and some of it printed at Wilkin's expense. Curtis would have completed it long before his *Guide* (1829) appeared 'had not one been promised from another quarter', a swipe at J. F. Stephens. Before the *Guide* and *British Entomology* there was thus no printed catalogue of insects, 'no regularity or harmony in nomenclature, and even in the boasted Cabinets of that time many of the largest families were '*rudis indigestaque moles*'.*

Curtis then describes how the work was done and presented, and here we have added some extra items in order to complete the picture. Four plates were supplied on the first day of each month, together with two pages of text describing the general represented. The actual plates themselves measured 20 by 11·6 cm., had rounded corners and were etched. The pulls from them were about 24 by 15 cm. Only bound sets are found today and their size depends on the amount of trimming done by the bookbinder. My own set has pulls measuring 22·8 by 13·7 cm. Sets could be obtained plain or coloured, costing 3s. 6d. and 4s. 6d. respectively. From 1826 it was announced that only coloured copies were to be supplied, but some plain pulls of 1828 have been observed, so possibly the plain series was reintroduced.

The plates were numbered as they appeared, the number placed at the top, and they were signed and dated at the bottom.

* A clumsy, unordered mass.

Hand colouring was used. There was no system in the selection of the successive titles; Curtis just used whatever he had to hand and whatever he thought would particularly appeal to his cash customers. A moment's reflection will show, of course, that it would have been impossible to do otherwise. For instance, if he had had to wait until he had got drawings of all the *Coleoptera* together before starting on the beetles, the work would have been much delayed. Curtis does point out, however, that this necessity added to his difficulties by increasing work. For instance, he might find himself dissecting, drawing and describing two similar insects at an interval of a year or more. This meant he pretty well had to start again from scratch each time, whereas if they had followed each other the work on the first would have helped the second. An example is two of the 'humming bird moths', No. 40, the narrow-bordered Bee Sesia (*Sesia bombyliformis*) was dated 1 October 1824 and the humming bird sphinx, No. 747 (*Macroglossa stellatarum*) carries 1 July 1839, fourteen years and three months later. Perhaps this wide spacing may have had a certain 'cliffhanger' quality about it, whetting the appetite for the next exciting instalment, but fourteen years does seem rather a long cliffhang.

In all the work ran to 193 issues with 770 plates. The separate parts can be bound up in one of two ways, firstly as issued, in which case the plates run from 1–770 and are a haphazard collection and usually sixteen volumes are made of them, secondly it can be bound arranged systematically, that is all the *coleoptera*, the *lepidoptera* and so on together, when eight volumes result. Curtis provided title pages and elaborate indices for the purpose of finding the way around this complex series. Obviously the systematic method of binding is far more useful and most copies found today are arranged in that way. It is slightly more difficult to find any particular insect, because the pages are not numbered consecutively. However, an index of running titles at the end of each volume does show one whether the item sought is at the start, middle or end of the book. One locates a number more or less at random and then moves forwards or backwards to find the desired insect. The whole thing, of course, is a bookbinder's nightmare and it is surprising how seldom sets are misbound.

In addition to the adult insect shown on each plate there were also finely etched details of distinguishing parts of its anatomy and the plant on which commonly it fed. The addition of the plant was important on two counts: it instructed the reader as to where the creature might be found, but it also enabled a very fine, balanced, artistic plate to be prepared. Curtis received praise everywhere for the accuracy and beauty of his plates. The two pages of text with each plate give the following information: the scientific and English (if any) names, the order, family and type of the genus, with its author (usually Linnaeus or Fabricius), a description of the genus, then of the species illustrated and its precedence. This is followed by running notes on the creature, and related species, such as the meaning of the name, its habits, possibly its life history, where it has been found, and particularly who found it, the plant or plants it feeds on, anecdotes about it and so on. Finally a note on the plant illustrated is given. Plates and texts combined are thus of great value. Although Curtis undertook to deliver four plates with each number, the first two sets had five each, made up as seen on p. 36.

In this table 'cabinet of' refers to the particular collection Curtis used for his main type specimen. It will be noted that his subsequent rival, J. F. Stephens, figures twice. Dale is the man who encouraged Curtis to come to London and to launch *British Entomology*, probably with financial aid. N. A. Vigors was a keen amateur zoologist, being particularly interested in birds and insects. He was a wealthy Irishman and was born in 1785 at Old Leighton, educated at Trinity College, Oxford and Lincoln's Inn; he purchased a commission in the Grenadier Guards, served in the Peninsula, was badly wounded at Barossa (5 March 1811) and returned to England, where, invalided out of the army, he began to collect birds and insects, eventually forming an extensive collection. He thus became a customer of John Curtis. In 1826 Vigors was appointed the first secretary of the new Zoological Society and presented it with his collection of birds and insects. Mr Vigors was active in the present Entomological Society in its early days, that is in the 1830s. He was an advanced Liberal, seldom spoke, never married and lived in Chester Terrace, Regent's Park, London, where he died on 26 October 1840.

Of Messrs Raddon and Bentley we know very little; the former was a fellow of the Entomological Society in 1848. Obviously both were collectors of insects.

If we now turn back to the ten insects used in the first two deliveries of *British Entomology* we can see that they set the tone of the whole. It is essentially a collector's work. The insects chosen are comparatively rare and the two moths (Plates 2 and 7) are big and spectacular. In fact throughout the whole series there are no more than a couple of dozen common insects, such as the red ant (752), hive bee (769), bed bug (569), horse fly (525), cockchafer (406), earwig (560), gnat (637), ladybird

Plate No	*Insect*	*Order*	*Cabinet*	*Plant*	*Vol No in systemic arrangement and date*
	First delivery				
1	*Cincindela sylvicola*	Coleoptera	J. F. Stephens	Meadow grass	I, 1.1.24
2	*Velia rivularum*	Hemiptera	Dale and JC	Rush	VIII, do
3	*Deilephila euphorbiae* (Spotted elephant hawk moth)	Lepidoptera	Raddon and JC	Euphorbia	V, do
4	*Petastes pini*	Hymenoptera	Bentley	Spruce	III, do
5	*Ctenophora ornata*	Diptera	Dale	Common knot grass	VIII, do
	Second delivery				
6	*Nebria livida*	Coleoptera	Vigors	Hop trefoil	I, 1.2.24
7	*Odenesis pini*	Lepidoptera	Brit. Mus.	Scotch pine	V, do
8	*Chrysis fulgida*	Hymenoptera	J. F. Stephens	Shepherd's purse	IV, do
9	*Anthrax ornata*	Diptera	Dale, Bentley and JC	Common chamomile	VIII, do
10	*Notonecta maculata* (Spotted boat-fly or water-boatman)	Hemiptera	Dale and JC	Marsh mousear	VII, do

(208), mosquito (537), cinnabar moth (499), mayfly (484), raspberry beetle (618), wasp (760), click beetle (694), apple-blossom weevil (562), turnip fly (630) and a few others. This may be considered a sensible move; with such high production costs there is not much point in illustrating common insects.

A list of subscribers was added to the last issue which is of great interest. It is very socially conscious. The list is in alphabetical order but the titled, at least those of a certain rank, precede mere mortals under each letter. Under A. the Earl of Aylesford heads the list although alphabetically he would be the last A. Why Lady Cox is placed alphabetically and not in the front with Lord Carteret it is difficult to see; perhaps she was a mere knight's widow. It was not male chauvinism, for Lady Drummond heads the D's and the Marchioness of Hastings her letter. Rank counted among the aristocratic too; the Earl of Sheffield immediately precedes Sir John Sebright, Bart. However, the most intriguing and difficult distinction is among the common run of men – the Misters and the Esquires. How was the decision made? Why is R. Wood of Manchester Mr whereas G. W. Wood, of the same city, is Esquire? Can Mr B. Clark, FLS of the Institute of Science, Paris, ever hope to become Esquire? What problems there must have been. In all there are 272 names, of which 34 were ladies. Of the original 272 there were 55 drop outs, that is people who had not paid up and for whom back numbers were no longer held. There was an additional list of 87 names for whom copies needed to complete sets were being reserved. There were thus 130 firm subscribers, but the work was so beautiful, talked of and wanted that we feel these 87 must have taken up their options to complete their sets. This gives us an idea of how much Curtis got for the first edition of *British Entomology*. The set was of 193 issues, at 4s. 6d. each coloured and 3s. 6d. plain. The demand was for coloured sets so we can take the price of a complete set as 4s. 6d. by 193 = £43. 4s. 217 sets at £43 4s. is £9,417 8s., a considerable sum for those days. In addition to this list of subscribers there would also have been sales made over-the-counter and from bookshops. But this is to neglect the lapse of time and the expenses, which were considerable. The £10,000 plus was earned over fifteen years, that is some £700 per year, and the plates had to be made, pulls taken and coloured and letterpress set and printed. Curtis says it cost him more than £3,000 to colour the plates alone. This sounds a great deal of money. The subscribers would need 16,524 pulls and additional sales could make this a total of 20,000; this number of plates costing £3,000 gives us a figure of 3s. each, instead of the current going rate of 1d. Admittedly Curtis's colouring is particularly good, so he may well have paid rather more. Also, perhaps, he counted in the cost of the studio, equipment, paints and brushes in the manner of the modern author making up his tax return.

Curtis went into the production of *British Entomology* with tremendous enthusiasm and energy and so too did J. C. Dale, who financed much of the work. The latter seems to have lost his enthusiasm for the publication rather sooner than Curtis. Writing from Glanville Wooton, near Sherborne to Haliday in Belfast, 6 February 1833, Dale says:

I see Ellis has Curtis's book at a reduced price now and I fear C. has made but a sorry affair of his speculation – as *I* always feared he would, but he thought different then. . . .

The letter, we may observe in passing, continues on an interesting modern note. How the countryside is being ruined by progress:

Wittlesea Mere has lost its character and I never probably shall try it again – crops of 'waving corn' and turnips now occupy the formerly sedgy edges of the lake which is nearly drained. . . .

A necessary adjunct to *British Entomology* was Curtis's *Guide to the arrangement of British Insects*.[14] This was essentially the list of insects found in Britain that Curtis had been waiting for from Stephens, but which never appeared. Such lists were essential for a collector – the equivalent of the philatelist's stamp catalogue of today: it showed him where the gaps were and what was needed to complete his 'set' and moreover a collector could send a marked copy of the guide to fellow collectors showing what 'swops' were available for exchange and what his 'desiderata' were. The *Guide* could also be cut up and the names used as labels in gentlemen's cabinets: very good business for the author for at least three copies would have to be bought by a collector, two for cutting and one for reference; it is possible that there were some copies of the *Guide* printed on one side only, so that they might more readily be cut for labels. The list was classified essentially on the Latrielle, Fabricius system and Stephens naturally disagreed with it and eventually published his own catalogue.

The first edition of Curtis's *Guide* consisted of two columns of insect names on eight sheets and was sold at 1s. per sheet. There were some 15,000 insects named of which Curtis had about 6,600 in his own collection. Naturally it makes considerable reference to *British Entomology* and Curtis tells collectors:

Four cabinets therefore of twenty-four or twenty-five drawers, 18 inches wide and 17 inches long, will hold all the British Insects and anyone wishing to collect any portion only is advised to purchase a good cabinet containing twenty-four drawers.

In the second edition, 1837, Curtis is still appealing to the collector rather than to the scientist. The list is printed with marks to show the species present in Curtis's collection and he adds:

As it is only by collecting diligently, exchanging liberally and purchasing judiciously that a good collection can be formed, the author invites Entomologists to send those insects which are desiderata to him with a marked guide, and he will be happy to make the best return in his power.

Curtis offered to pay £2. 10s. per 100 for insects he did not possess. 6d. each was quite a considerable sum in those days and is indicative of the interest in collecting. Some enormous prices were paid for rare and spectacular insects. An example is *Mormolyce phyllodes*, a large beetle from Malaya which has leaf-like elytra. In 1832 Curtis tells Haliday he had obtained a pair, 'for a specimen of which £40 was recently given. [Note: the writing is faded and the figure may be £20], but he says he must part with one as he cannot afford to keep both.' Obviously the collecting disease could be very strong. It persisted and was widespread; for instance M. J. Desbrochers des Loges, from Tours ran a flourishing business in Coleoptera in the 1980s. He issued regular price-lists and ran a journal *Le Frelon* (*The Hornet*) a curious name for a paper devoted entirely to beetles. In these lists prices run from 12 francs (9s. 7d.) for *Eutrachelus Temmincki*, male, to 15 centimes (1½d.) for *Phyllotreta brassicae* – the flea beetle pest of turnip and other brassica crops. Who the big spenders were who purchased such a common insect is difficult to visualize. The collector's world was so divorced from any practical application that many would not know it was a pest and others who might suffer from the pest would not recognize it under its Latin name. It is surprising that more pests were not spread around the world by these dealers as many especially bred exotic insects for their customers. M. Desbrochers des Loges advertised the 'Anthonome d'Amérique', presumably the cotton boll weevil, but did eschew the 'Doryphore' (*Leptinotarsa decemlineata*) or Colorado beetle potato pest, which some English dealers (not Curtis, he had too much sense to do such a thing) were breeding in order to supply this pretty, but dangerous insect to their customers. The first quarantine legislation in Britain was designed especially to stop this practice.

In 1829, after five years of publication, Curtis began to find his edition was too small to meet the demand and he started to reprint previous parts, that is Plates 1–30. Mr R. E. Blackwelder[5] in 1947 wrote an interesting paper on the various editions of Curtis's work to which readers interested in the details should refer. Curtis also took advantage of reissues to make, often considerable, corrections. In general the reissues of both plates and texts bear the original date, but, with the exception of Plate 30, and those after Plate 34, the new edition has a line under the number. An example is Plate 11 (1 March 1824); the original *Molorchus minor*, Fabr, a beetle, has been changed (1834) to *Necydalis minor* and the text rewritten.

6 *The Rivals*

In 1826 a cloud no bigger than a man's hand appeared on Curtis's horizon: it was the proposed production of a rival publication – *The Illustrations to British Entomology* by J. F. Stephens. Stephens was a gentleman,* son of a naval captain, had a sinecure job in the Admiralty and spent his time classifying insects for the British Museum. He knew the right people. He should never have plagiarized Curtis's title, and what made matters worse for the latter was that John's younger brother, Charles Morgan, made most of the drawings for Stephens's series. Of the twenty-four plates in the first two volumes only two were not drawn by Charles Morgan, the exceptions being done by the author himself and the young entomologist, I. O. Westwood. Stephens did two of the Pierids, the large white, then called *Pontia chariclea* by Stephens and the small white, which he called *P. rapae*. Stephens appeared to think that his *P. chariclea* was different from the commoner *P. brassicae;* actually it is more likely that he was merely confusing the spring and summer broods of the large white which tend to be of different sizes and markings. The large and small white are the two insects which can be such severe pests of cabbages and other brassicae, both then and now, and Stephens's account is an interesting example of what may be called the 'collective' attitude of the age (although it is giving the word a new meaning). He was only writing for collectors, arguing about names, varieties and classification. Although he mentions that these two (or three) species can be 'found in abundance' and that 'in the year 1818, when the various species of *Pontia* abounded so greatly near the metropolis as to attract the notice of the public journalists,' yet he never considers them as a pest of farmers' or gardeners' crops, nor has he the slightest reference to the damage they do. Aiding agriculture was *not* the task of a gentleman scientist. Curtis, on the other hand, has *Pontia* as No. 48, 1 December 1824, in the halcyon

* As the suspense may be unbearable, let me quickly add that Curtis eventually became a gentleman.

days before the Stephens threat; he thus is more concerned with *Pontia* as a genus (we now call it *Pieris*) and he illustrates *P. Daplidice*, the Bath white, now a rare insect and a migrant from Europe. Unlike Stephens, Curtis showed some interest in the economic aspect of his science, even at this early stage. Of *Pontia* he says it 'contains five British species, which, with the exception of the one figured, are amongst the most common of our butterflies, the caterpillars of many of them being the greatest pests amongst our vegetables, by feeding upon and destroying the different varieties of cabbages cultivated in our kitchen gardens'. But the collector is stronger in him than the economist or scientist. He illustrates the rare Bath white; it has more customer appeal than the other common whites.

The Charles Morgan drawings for Stephens were mostly engraved by George Wagstaff, who used lithography. Each plate usually has three or four insects on it; they are not as well executed as Curtis's; there are no enlargements of vital distinguishing parts, nor are there any plants, which meant that the plates were not as useful as those of Curtis, nor were they as attractive as works of art. Charles Morgan was not as good an artist as his elder brother, nor was George Wagstaff as skilled an engraver.

Nevertheless, Curtis found the whole thing very worrying and it led him to alter his plan for *British Entomology*. It will be remembered that he set out to describe the *genera;* when he heard about the competition from Stephens he tried to meet it by describing as well as many *species* as possible. The change begins about Plate 150. Before this number the main object is the genus; afterwards a list is given of the species known to occur in England, who found them and where they can be found, but as the letterpress is still confined to two printed pages there is no room to describe the species at length, and even information about the genus itself had to be curtailed, though more was incorporated by using a smaller type face. Some examples should make this clear. Stephens's first plate appeared on 1 June 1827, nevertheless Curtis had considerable notice of the forthcoming publication. A pre-storm Curtis plate, such as No. 120, *Alomya victor*, an ichneumon fly, concentrates on the genus, with a brief mention of another species (*deballator*) at the end and is dated June 1 1826, but by November of the same year Plate No. 141, *Alysia apii*, another ichneumon, has a list of nine species given, with brief identification notes.* Later these notes tend to turn to mere lists of names and places. For instance, Plate No. 183, a beetle *Agonum austriacum* Fab., October 1, 1827, has a list of nineteen species on the second page of the letterpress; none of these are described, though they are divided into two classes, 'Thorax suborbicular' and 'Thorax somewhat obovate'. The necessarily brief notes there are are concerned with the place at which specimens were found and who found them. Curtis also advises his subscribers that all but one may be seen in his own cabinet. This, no doubt, would lead enthusiastic collectors to his house and perhaps lead to more business being done. This was another instance where the Victorian zest for competition and *laissez-faire* economics was acting against the best interests of public and protaganists. Competition among the rival entomologists meant that neither of them did as well as they might have done working alone, or better still in partnership with each other. The same thing occurred, on a much larger scale, with railways: competing lines tended to be built, such as those from London to Bradford and from London to Exeter, almost parallel – an inefficient use of resources both for the shareholders and the public. In the case of the rival entomologists there was a kind of Gresham's law of effort; both had to direct their respective publications to appeal to the rich collectors with a resulting loss to science.

To begin with Stephens and Curtis were fellow entomologists and friends. In 1824 the first insect illustrated in Curtis's *British Entomology* was from Stephens's 'cabinet'; it is a greeen beetle with red spots on the thorax and yellow ones on the eleytra – *Cincindela sylvicola* (Plate 1, Vol. I systematic). Curtis writes, 'I am indebted to my friend J. F. Stephens Esq. for being able to present my readers with a figure of this rare and elegant species'. In No. 4 the ichneumon *Peltastes pini*, Stephens is mentioned as having bred this insect from a lobster moth caterpillar. In the second delivery (Plates 5 to 10) Curtis again uses Stephens's cabinet with Plate 8, *Chrysis fulgida*, a Hymenoptera, although it was collected by a Mr Standish whom Curtis also publicizes at this stage; in a footnote he has 'Mr Standish, 10 Prior Place, East Lane, Walworth; who always has insects of his own

* An interesting error, not previously noticed, as far as I am aware, can be found in the letterpress of No. 141; this insect has been put in the order Coleoptera instead of Hymenoptera, no doubt a misprint which much pained Curtis.

collecting for sale.' Stephens probably bought the specimen from Standish. Obviously there was a certain amount of professional solidarity in the trade in the early days of *British Entomology*. Whether it is significant that Stephens drops in the personal and social scale from Plate 1 'my friend J. F. Stephens' to 'Mr Stephens' in Plates 5 to 10 is difficult to say; Curtis may well have had some more or less unconscious forebodings.

In any case this happy state of affairs slowly began to pass away and after a while Curtis began to see a villain behind every bush. He had much provocation, for example the imitation of his title and his brother's making of the drawings for his rival, but the truth of the matter is that the project was a most expensive one and there was not the market for two such similar publications. There was hardly a market for one, unless subsidized or under patronage, for the purchasers of expensive works were more collectors than the scientists Curtis wanted to appeal to. Their motto was more 'What a glorious day. Let's go out and kill something' than 'What is the history of this insect and how is it related to that one?'

Curtis, in his early enthusiasm, had underestimated his costs and difficulties. In addition his scientific mind was annoyed by the errors Stephens was making and at his arrogance in presenting his material. Stephens on the other hand saw himself as the apostle of the new nomenclature and was aware that the Curtis plates were superior; thus he began to use any stick that came to hand with which to beat his rival, the nearest being to dispute his rival's method of classification, for there was no agreed system, as we have already noted. Curtis soon began to see the entomological world as made up of two camps – pro-Stephens and pro-Curtis, as can be seen from various letters to Haliday, his friend in Belfast. They increase in acidity as time goes on and Curtis's financial difficulties become more pressing:

19 December 1830: It gives me great pleasure to find that you cheerfully give me credit for the pains I take with my work, I regret to find that Stephens, Mr Hope and the Reverend Mr Rudd hold a very different course; they both say they are the best friends I have, one boasts he has induced twenty people to discontinue it (observe he is a man of the greatest veracity) and the other says he takes it to encourage me, but it is quite wasted paper. I do not envy either.

The Reverend G. T. Rudd was one of the founders of the present Royal Entomological Society and a considerable collector. In 1833 he wrote that his

mansion in Yorkshire was totally destroyed by fire; in it perished my collection of insects which had been gathered after many years of exertion. . . . I am at once reduced to a kind of entomological pauperism,

and he asks his correspondent (Haliday) if he can send a few insects 'through our friend Mr Curtis'. In effect Haliday and Curtis and possibly many others felt sympathy for the distressed entomologist and sent him insects with which to re-establish his collection. It seems to have been no ordinary gift for on 11 February 1837 Rudd thanks Haliday, though in a somewhat roundabout way, he hopes that:

Our friend Mr Curtis has *long 'ere* this expressed to you my very cordial thanks for the very liberal and acceptable contribution you so handsomely forwarded.

Why he had not thanked Haliday directly it is difficult to see. Rudd became a villain, as we shall see below, but first we must investigate Mr Hope. The man to whom Curtis refers could have been one of three people, either Mr T. H. Hope, a fellow of the Entomological Society in 1863, or Mr S. T. Hope (1837) or the Reverend Frederick William Hope (1797–1862), another of the scientific clergy. Most probably it was the latter. The cloth, in Curtis's view, in no way conferred entomological purity. F. W. Hope was a curate at Frodesley, Shropshire, but 'being possessed of ample means formed a large collection of insects and invertebrates'.[53] He was much concerned with the formation of the Entomological Society (1833) and was its first treasurer; he presented the society with many books, the foundation of its library. He gave his collection of insects and invertebrates to Oxford University in 1849 and appointed I. O. Westwood its keeper; he also established the Hope Professorship and nominated his protégé (Westwood) to the chair, but this was much later (1861), a year before Curtis's death. Altogether he was a distinguished man (FRS, Doctor of Civil Law, etc.) and it is a pity that Curtis thought of him as in the Stephens camp. One feels though that with all this entomological activity his parish must have suffered for he was but a curate: perhaps he gave it up at an early stage.

27 October 1832: Westwood and Stephens are:

Hand in hand and will do all they can to render my guide useless, which I shall be very sorry for, I think you had better apply any of my specific names to species that answer them but the Genera are the most important because they have gone abroad to my friends both here and on the continent, unfortunately Westwood is a great favourite in Paris, his impudence having astonished the French. He carries everything before him, even Dejeune unfastened his boxes that were supposed to be hermetically sealed for this illogical and vain fellow, such magic is there in assurance.

We may note in passing the ambivalent nature of the word 'fellow'. To be a fellow of the Royal Society was (and is) the highest intellectual honour; on the other hand Westwood was a 'vain fellow'. One is reminded of Mr Pickwick, just before Mrs Leo Hunter's fancy dress breakfast, quarrelling with Mr Tupman on the matter of the costume the latter was proposing to wear to this event. The dispute reached a peak of fury and drove Mr Tupman to say to his President, 'Sir, you're a fellow!' and for Mr Pickwick to reply, 'Sir, you're another': a dreadful state of affairs for the club, only solved by the intervention of Mr Snodgrass.[25]

On the 23 December 1832 Curtis is referring to 'that pert lad Westwood' who was, in fact, twenty-seven years old at the time.

13 February 1833: Curtis's difficulties are pressing him and he is getting critical of many other entomologists. Edward Newman, who was editor of the *Entomological Magazine*, which ran for five volumes (1826–32), comes under Curtis's fire:

Like many system-makers Mr Newman has assumed a great deal, which, like mortar in a stone wall, may make the most heterogenous masses unite until the true labourer comes when a single blow will knock it to pieces and show all its deformity.

Curtis continues saying he respects Newman and regrets that he (Newman) is writing a learned book with very little knowledge,[55]

and with a tolerably good opinion of his work and a very moderate one of others, as is evident when he says 'no one has *thought* on the subject', excepting himself anyway – If he be not blinded by vanity and pursues his researches with the vanity he possesses he will in a few years repent of his *Sphinx vespiformis*, at least such is my opinion.

Newman's book was *The Grammar of Entomology*. The dispute, as usual, was about naming and presumably Newman did repent about his *Sphinx vespiformis* for in a later edition of his book (1841) he points this out and says he was trying to show 'it was synonymous with *Aegeria Asiliformis* of our cabinets'. Curtis called this creature *Aegeria ichneumiformis* and it did resemble an ichneumon fly. In both their eyes it had the merit of being rare! The fact was that, with all this vast collecting enthusiasm in being, new insects were being discovered all over the place and enthusiastically named by their finders. No wonder the reckless creation of new genera, species and even families created confusion. There was no central body to sit in judgement on the matter and, though there is today, the scientific names of insects still swivel and change and the same sort of taxonomic disputes arise, though with a little more order about them now and, we hope, less hard feelings. In the course of my lifetime I have seen the apple pest – the codling moth – pass from *Carpocapsa pomenella* to *Cydia pomenella* and come to rest at the moment as *Ernarmonia pomenella*, all Linnean names, it having finally been decided that *Ernarmonia* was the best genus for this species. The working entomologist may consider himself fortunate that the specific name has remained constant, as is usual in such cases.

Curtis also found the Reverend Doctor G. I. Kidd another suspect cleric; in the same letter mentioned above, which was a long one, Curtis continues:

> I also send you a letter from the Revd G. I.* Kidd he may take advantage of this to insinuate that he is my friend, I must explain. He was and still is I believe the intimate friend of Messrs Hope, Stephens and Westwood, he once behaved in my chambers so Cavalierly that I did not speak to him for some time and he has lately treated Mr Davis in an equally ungentlemanly manner.

Curtis continues by saying that he thinks the reverend gentleman stole specimens from his (Curtis's) collection and gave them to Stephens, for whom he is a 'spy'. He also recounts that at Dale's table the Reverend Kidd decried Curtis's *British Entomology* and so praised Stephens's *Illustrations* that some of the guests said they would give up taking Curtis's work. The host Dale, however, defended Curtis and said 'I wish Curtis were here to speak for

* The writing is indistinct; the initials may be G. T.

himself' and at this moment Curtis came into the room, a situation causing considerable embarrassment all round!

By the autumn of 1833 things were a little brighter for Curtis for on the 10 September he is able to tell Haliday that 'Mr Kidd, I am happy to say has got a living which has carried him to the North'. Although one of the gad flies had been removed others remained:

24 February 1834: Curtis reports that while Dale and others were defending him:

Newman and his band are apparently daily sinking in estimation,

and of Mr F. Walker (who was to write three volumes on Diptera) he says:

I fear Newman has wound him round his finger.

The disputes about names continue, as they had to under the circumstances. The following year we find:

22 February 1835: Curtis thanks Haliday for his continued praise of his work and continues:

I think you would be doing much good and stopping a villainous and most unjust course pursued by Westwood and adopted by Stephens, if when you had an evening to spare you would think of the subject and send your remarks to the *Entomological Magazine*. I allude to their substituting their own names for those of the original writer wherever they can find that the name has been employed in any other branch of science and as they are sworn friends and Westwood I understand has a list of Genera they are able in this underhand way to raise themselves on the shoulders of others. As I have no affection for one and everybody knows how much I despise the other I dare not trust myself to enter on the subject.

In 1836 The Reverend Doctor Rudd, in spite of having returned to his Yorkshire home, is still worrying Curtis. The latter, after discussing ways of increasing his own income (selling his duplicates to Dublin or to English residents abroad) says:

That illiberal humbug Rudd has been with me and left a letter for you which I enclose, he said he had taken the *Paururus* in plenty and wanted to see mine; also several males of my *Alyson* with . . . [name illegible, possibly 'Skrimshire'], one of which Rudd had not yet disposed of, yet Rudd is the man who begged me to assist him when his cabinet was burnt and I sent him eighty named species, *not rubbish*, and Dale and I afterwards sent him a small box of very rare insects.

Obviously Curtis thought Rudd should have given him his spare *Paururus* and common gratitude suggests it would have been a proper gesture, but possibly the clergyman was giving Curtis a lesson in his professional capacity on the commandment 'Thou shalt not covet thy neighbour's Hymenoptera'. And so the struggle continues. We may note that Curtis and Stephens only once both illustrated the same insect, namely the ground lackey moth *Clisiocampa castrensis*. Both entomologists used the same name, the genus supplied by Curtis and the species by Linnaeus. Curtis, as always, has a much better plate, shows the moth, both male and female, the antennae and mouth parts, caterpillar and main food plant: Stephens, as nearly always, shows the moth alone and then only the male and the plate has two other species of moths on it. Curtis not only makes a much better picture but has a far more accurate one as regards shape and colouring. Stephens's plate is dated 1 July 1828 and Curtis's 1 September 1828, but must have been prepared well in advance of the appearance of the former. This lackey moth then was the only coincidence between the rivals. Was this by chance or did an arrangement exist between them not both to do the same insect? If this latter, Charles Morgan, John's younger brother, who did most of the drawings for Stephens, was likely to have been the go-between. The figures rather suggest there was some agreement and that the lackey moth coincidence was an oversight. There were 1,900 Lepidoptera known: Curtis illustrated 199 and Stephens 119. A deliberate overlap was thus not very probable and the single one we find was more likely to be due to chance than anything else. The point is that if they had both ignored each other we could expect to see about a dozen the same in both sets; but if each of them avoided doing any *Lepidoptera* done by the other, as the series proceeded, we would expect to see few or no overlaps, as is actually the case (see Appendix 3). The respective publications obviously led to much comparison and discussion by the interested, surely to the advantage of Curtis's work.

Looked at in today's light it is difficult to see what Curtis was afraid of as his publication is so self-evidently the better. English judgements of his day were perhaps clouded by local faction fights over naming; one was either for or against Curtis. The quarrel undoubtedly was much discussed. The *Entomological Magazine* used to start off its numbers with what today seems to be a body of somewhat heavy humour, but no doubt was most witty then: it was rather in the style of *The Compleat Angler*. In a *Colloquia Entomologica* of 1835 the Curtis/Stephens affair is discussed, probably in an attempt to pour oil on the troubled waters. A short quotation from it will not come amiss (short, because it goes on for twelve printed pages, twelve pages of joking presumably being held to be twelve times as funny as one page).

Venator:	Success to the study of entomology.
Omnes:	Success . . . ditto.
Venator:	I would from courtesy exclude the present company, otherwise a reply would be requisite. Of those not present Kirby, Leach and MacLeay seem to be the most eminent.
Entomophilus:	Decidedly. But the accurate pencil of Curtis!
Erro:	Which?
Entomophilus:	The elder brother, John Curtis: his *British Entomology* is beyond all comparison, the most valuable work on Entomology our country has produced; and I consider it far from being a credit to us that its circulation is so limited.
Erro:	His brother is a good artist: his drawings in Mr Stephens's early numbers are very beautiful: equal to any I know of.
Entomophilus:	We must not refuse our meed of praise to Stephens; his undertaking is a most laborious one.
Ambus:	Yes: I hope he will have courage to repeat it.

They proceed to draw lots for the post of editor; (*an indication that the popularity of such a post was as low then as now*) *: his troubles are described:*

An Editor leads an easy life.
The pleasures, believe me, are many –
Plague, jealousy, envy, fear and strife.
The evils – I never found any.

Naturally this attempt at pacification only succeeded in infuriating both parties, more particularly Stephens as he was so much more easily infuriated. In fact these joking introductions were soon dropped as Stephens threatened dire things, even recourse to the law.

Even the Reverend Mr Rudd, whom Curtis so suspected, was not exactly pro-Stephens; on the difficulties of describing species he writes:

I am afraid Mr Stephens affords just ground for apprehension, that to describe 300 insects per month is a task more easily undertaken than well accomplished. However *Nous verrons.*

Curtis was sorely tried and expresses his feelings in a letter to Haliday:

13 February, 1833: I am the last in the world to embarrass Science by petty factions or to dictate to my friends, but as men behave so will I speak of them . . . bitter enemy . . . hypocrisy etc.

The above might suggest that Curtis was an irascible man always looking for injuries, but I do not think this is the case, as we shall find out when we come to discuss his private life (what little is known of it) and his general character. Moreover the evidence for the injuries done to Curtis does not depend entirely on the letters from Curtis to Haliday. An entomologist named Hooperhoe was commissioned to write the article on John Curtis for the *Dictionary of National Biography* at the turn of the century. The Hooperhoe papers are in the Norwich Records Office and contain many letters and notes. Hooperhoe took his task very seriously and entered into extensive correspondence, although, as has been mentioned, he was wrong about Charles Morgan being the elder brother.

Among the letters is one from C. W. Dale, the son of J. C. Dale, Curtis's friend and benefactor. He writes:

Both Curtis and Stephens did much to promote entomology and it is to be regretted that J. F. Stephens had it in his power to do J. Curtis a great wrong and injury, both in retarding the sale of his work and in getting people to believe he was a bad and dishonourable man.[35]

The young Dale points out that Stephens was a quarrelsome man and with the same sort of hindsight we now have, but less distant then, pointed out that in the quarrels on names and origins they both could be mistaken, as, of course, is only too likely. Dale gives instances; here is an example – Curtis said the scarce Swallowtail butterfly (*Papilio podolarius*) and Stephens that *Colias Philodice* were British and both were wrong. Dale *père*, who had property in Hampshire, got Curtis an introduction to the Earl of Malmesbury and his brother Charles Harris, later Bishop of Gibraltar. This much helped forward Curtis's work, but it may also have given Westwood an opportunity to spread a spiteful tale, and we shall come to this again at the end of this chapter.

Curtis's basic difficulty was that he was a poor man trying to serve and indulge in the dilettante pursuits of the rich collectors on the one hand and the rich, or endowed, naturalists on the other. Moreover he, John Curtis, the son of a sign painter, seemed to have pretensions to gentility. One might smile at the efforts of a rich merchant and, of course, eventually admit him to the foot at least of the social ladder, but a scientist and insect specialist . . . really, there are limits to indulgence. One feels that the assured social position of Stephens, the Reverend F. W. Hope and most of the others, must have been a considerable contributory cause of Curtis's malaise, though the main one was his need for money in order to pursue science. His attempt to do it with *British Entomology* was not very successful, the work was continuous and exhausting. The scientist, like Orwell's thin man, imprisoned within the collector's slave and servant, was always trying to get out and could not, producing a vast frustration.

However, in spite of his success and social position, Stephens was not immune from quarrels, as we shall see. We have already mentioned that Kirby and Spence's great work led to a flood of entomological books in the mid-nineteenth century. Among them was a work entitled *A Conspectus of British Butterflies and Moths in Great Britain*, 1832, by James Rennie, much of which, Stephens maintained, had been taken from his *Illustrations of British Entomology*. Stephens brought a case against Rennie for infringement of copyright; it came before the Lord Chancellor, 19 July 1832, and developed into something of a legal picnic.[80]

James Rennie (1787–1867) was an able man; he was said to be the natural son of Thomas Rennie (or Rainey) of Aldenholme, Sorn, Ayrshire and Margaret Edwards. Like Curtis he had a hard time trying to make biological science earn him a living, but by 1830 he secured the professorship of Natural History at King's College, London. But the chair was abolished in 1834 owing to the lack of interest in the subject; there were no students. He then turned to literary work, writing comparatively cheap popular works for the new market already described (Chapter 3).

It turned out to be a hard and not very rewarding activity.

There were two books in 1830, *Insect Architecture* and *Insect Transformation;* in 1831 *Insect Miscellanies* and *Birds;* in 1832 *Alphabet of Insects* and the *Conspectus* . . ., the work that led to the legal action which so excited the entomological world. Rennie also wrote popular natural history articles for the press under the name 'Rusticus'. It is, of course, well known that things in the literary world (and many others too) moved much faster in the old days (to write and publish a technical book today takes at least two years) so that turning out all this stuff so quickly meant Rennie had to draw on the works of others and then 'popularize' them in order to live. The case against Rennie looked very black as Sir George Sugden, for the plaintiff, opened by pointing out that certain errors in Stephens's book had been copied *verbatim* into Rennie's, showing that he had not used the basic scientific sources. For instance, a printer's error in Stephens was 'siona' for 'scoria'; Rennie had used the former, erroneous word ('thus creating a class of insects which does not exist') and other instances were shown. The court, however, accepted the view that the two works were of quite different natures; plaintiff's was one of four volumes at one and a half guineas per volume and was intended to be of thirteen volumes – £19. 10s. in all; and defendant's was a popular work costing 7s. 6d.; one did not impinge on the other.[46] After much delay and legal argument the case was referred to arbitration and the arbitrators found for the defendant, leaving Stephens with enormous costs. The entomological world was most indignant at the verdict and raised a subscription to pay Stephens's costs.

The question now arises as to whether Curtis, an expert in this field, was called to give evidence; he would have regarded the case with mixed feelings. On the one hand he would be indignant at plagiarism, for much borrowing from *British Entomology* had taken place, and on the other he would have been delighted to see his own chief plagiarist (as he regarded it) discomfited and a poor and struggling scientist like himself defended. There is but one piece of evidence bearing on the case discovered so far, and that not very conclusive. In the *Supplement to the Illustrations of British Entomology* which appeared in 1846, when Stephens was in retirement in the country (Eltham Cottage, Foxley Road, Kennington), he is still sniping at Curtis. He points out that in the nineteen years from the first number of the *Illustrations* seventy-two works on entomology have appeared and that forty-one of them, that is more than half (57 per cent,) are arranged according to the Stephens's system. They were still rivals in this respect. The *Illustrations* had been brought to an abrupt end and the Rennie case still rankled with Stephens; he writes of the

abrupt termination of this work . . . cost of nearly £10,000 . . . loss of time in defending my copyright in 1832/3, my prospects were forever blighted by the inconsiderate prejudices of an implacable and tyrannical individual: thus compelling me in the decline of life*, of relieving myself of the thraldom of office, and of retiring in comparative penury therefrom.[77]

Who is this 'implacable and tyrannical individual'? It is hardly a description one would apply to the gentle and long-suffering Curtis or, under the circumstances, to Rennie, who, like the *méchant bête*, was merely defending himself. Perhaps Stephens saw Curtis in this way: the angle of view makes all the difference.

After this famous case Rennie became most unpopular in the profession, but continued to write, one of his most interesting books being the *Alphabet of Natural Theology*, but his lack of esteem and the poor returns obtained from scientific literature led him to emigrate to New South Wales in 1840 where he lived to the ripe old age of eighty.

We thus see John Curtis, James Stephens and James Rennie all competing with each other according to the classical Adam Smith to Marshall economic concepts, resulting in none of them producing as good work as they might have done, whereas if they had cooperated, say under a Keynesian economic system, they would have advanced biological science and the popular understanding of it at a much faster rate than actually was the case, and the classification of insects would have been achieved on an agreed basis that much sooner. Perhaps then Darwin might not have been such a traumatic shock. The energy and intellectual effort wasted by such competition was enormous.

We close this chapter with a few remarks on Westwood, to try and discover if he was such an adversary as Curtis thought him to be. John Obadiah Westwood was born in 1805 at Sheffield, the son of John Westwood a medal maker and die sinker. He was educated at the school of the Society of Friends. He was at first

* He was fifty-four.

articled to a solicitor and became a partner in the firm for a short while, but he preferred some profession where he could use his talents as an artist and so left the law (as did Curtis, it may be recalled). The young Westwood was highly skilled in drawing minute insects and in the production of illuminated MSS and, in fact, became one of the leading authorities on Anglo-Saxon and mediaeval manuscripts. We have already mentioned his association with the Reverend F. W. Hope and that he became the first Hope professor of entomology at Oxford. Westwood was an original member of the Entomological Society (1833), became president for three periods and at the Jubilee meeting (1883) was unanimously elected as life president; he regularly attended meetings and did much for the society. He was a lifelong opponent of Darwinism though he seems to have remained friends with Darwin because it was largely through the latter's influence that Westwood received the Gold Medal of the Royal Society.

Did Westwood spread the suggestion that Curtis was a natural son of the Earl of Malmesbury? It was by no means unusual in that age to explain any appearance of unusual talent in the lower orders by illegitimacy. What other reason could there be for it? Among the Hooperhoe papers, to which we have already referred, is a letter from J. C. Dale (21 December 1901) saying that Professor Westwood told him that John Curtis's father was Charles Curtis, an engraver and that this might account for both brothers being artistic. However, the professor then appears to contradict himself. The letter continues:

> The professor told me that Curtis may be a natural son of Lord Malmesbury, but my father never said so nor is there any proof of such. Westwood, you must remember was one of Stephens's party and was one of his artists.

The suggestion seems unlikely: moreover if he were a natural son of Lord Malmesbury he could not at the same time owe his artistic talent to Charles Morgan Curtis, the sign-painter. Curtis visited the earl about 1823 by which time the first earl had died and had been succeeded by his son (1778–1841). Now a man might welcome an illegitimate son, particularly a talented one, but hardly an illegitimate half-brother, so we are left with the suggestion of a spiteful story from Westwood. On the other hand Westwood wrote fulsome obituary notices of John Curtis in both the *Proceedings of the Linnean Society* and the *Annales de la Société entomologique de France* (the latter being much longer) and purchased from his widow Curtis's entomological diary, some MS and drawings, for the Entomological Society: then it is possible to think more kindly of a dead opponent than a living one.

The quarrel with Stephens eventually was healed. Dale (*fils*) records that the two men could be seen in friendly conversation at the British Museum, such is the power of insects, or of that great institution, or both together, to heal the direst wounds.

John Curtis, aged 59. President of the Entomological Society 1855 (by kind permission of the Royal Entomological Society)

BASEMENT

Curtis's four main residences. *Top*, now No. 57 Charlotte Street, London; No. 11 Robert Street, Hampstead, London, now pulled down; No. 19 Belitha Villas, Barnesbury Park, London, where he lived after his second marriage. *Bottom*, the Lower Close, Norwich, in the Cathedral precincts, where Curtis lived as a boy and where he caught his 'rare' specimen of the lobster moth

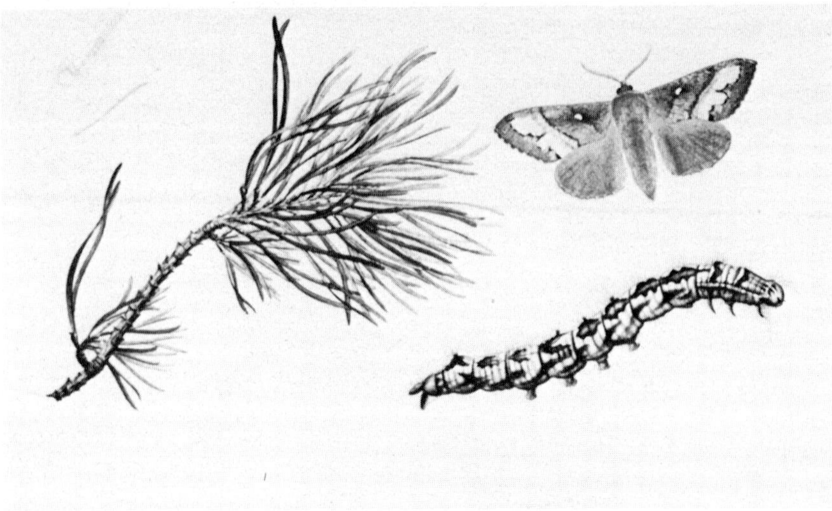

Drawing of the pine lappet moth, *Odenensis pini* Linn. with, *below*, a plate of the lappet prepared from the above drawing, showing a variation in the arrangement

Drawing of the strawberry weevil pest (*Otiorynchus Maurus* Gyll.) with, *below*, a plate of the above; the etching has exactly followed the line of the drawing

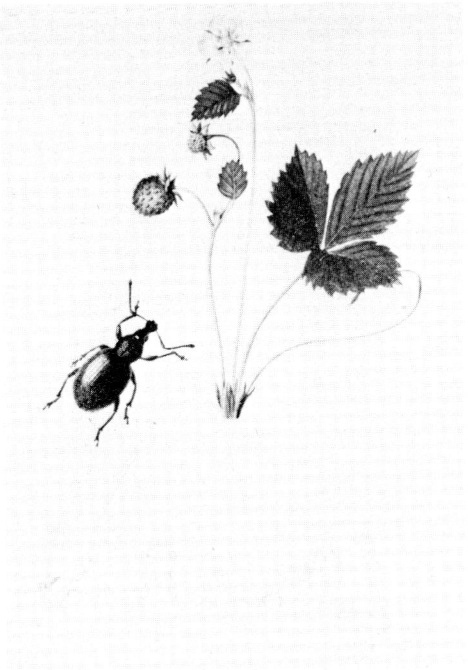

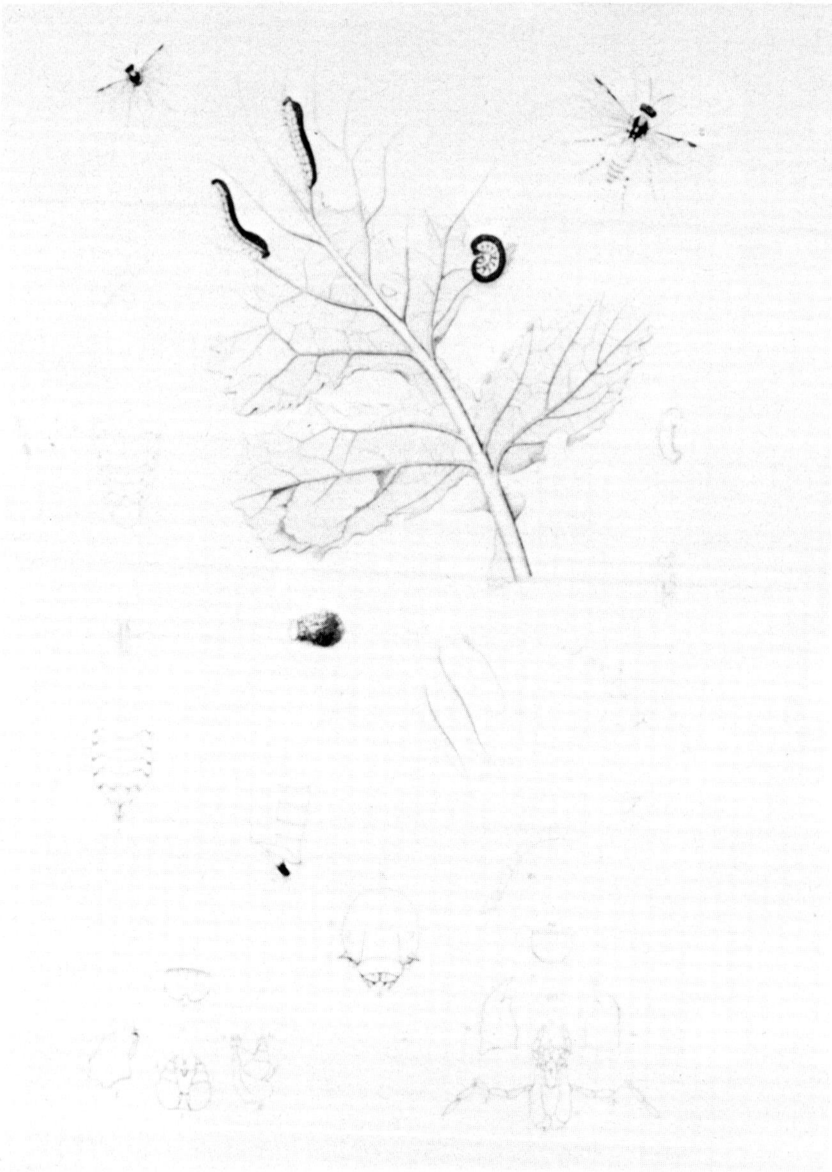

Plate B of *Farm Insects* showing the turnip sawfly (*Athalia spinarum*. Fab.) a severe pest in Curtis's day, but now no longer found

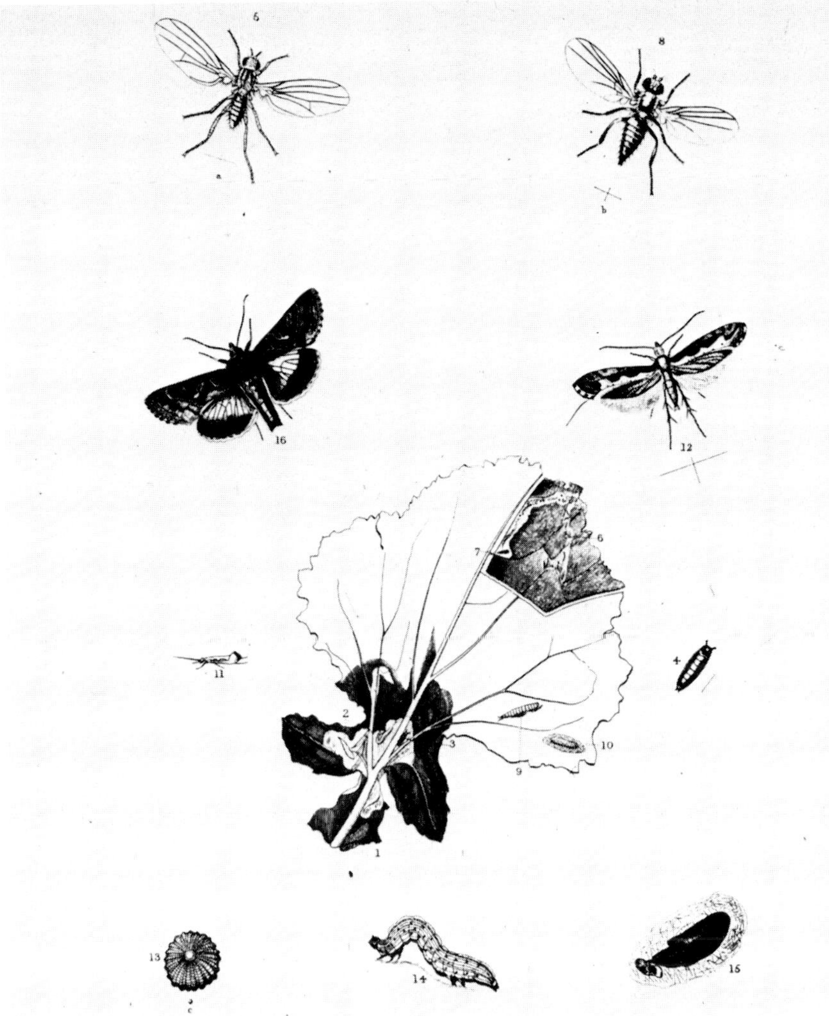

Plate C from *Farm Insects*. *Drosophila flava* and *Phytomyza nigricornis*. Two winged flies and moths whose caterpillars feed on turnip leaves. Drawn by John Curtis and engraved by J. H. Le Keux: 1. Turnip leaf; 2. Maggot of *Drosphila flava* feeding under cuticle; 3. Pupa of ditto; 4. The same, greatly magnified; 5. *D. flava* flying; 6. Gallery formed on the underside of a turnip leaf by the maggot of *Phytomyza nigricornis*; 7. The pupa of ditto; 8. *P. nigricornis* represented flying (larger than life) b. The natural dimensions; 9. Caterpillar of the turnip diamond-back moth; 10. Pupa of the same enclosed in the cocoon; 11. The moth at rest; 12. The same flying and magnified; 13. The egg of the Y moth; 14. Full grown caterpillar walking; 15. Chrysalis in its web; 16. The Y moth flying

Locusta Christii Curt. or Christy's locust found in the Clapham Road, London, July 1826; a very fine plate

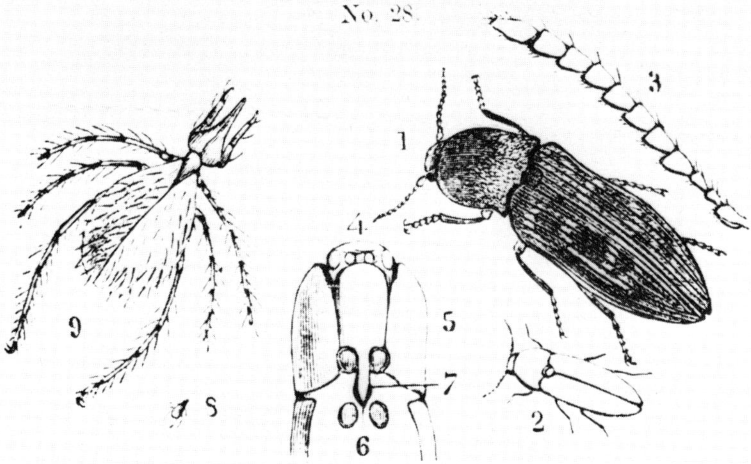

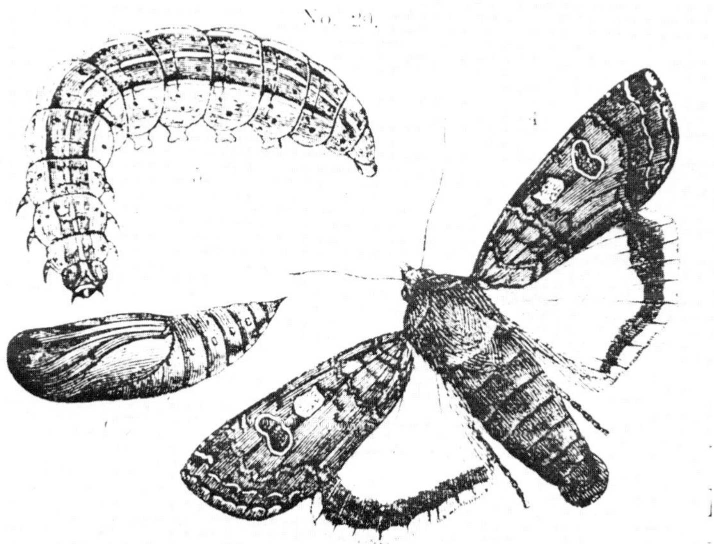

Above, Woodcut by John Curtis. Fig. No. 28 from *Farm Insects*, *Elater* (*lelidotus*) *holosericeus* Fab., the Satin-coated Click-beetle (wireworm); 1 and 2 the adult insect, 3 antenna, 4, 5 and 7 jumping apparatus, 6 base of body and four cavities in which legs were inserted, 8 and 9 a scarlet mite, *Leptus phalangi*, sometimes found on this insect

Below, Woodcut by John Curtis. Fig. No. 20 from *Farm Insects*, *Noctua* (*Triphaena*) *pronuba* L., the Great Yellow Underwing, or cutworm. Adult, caterpillar and pupa

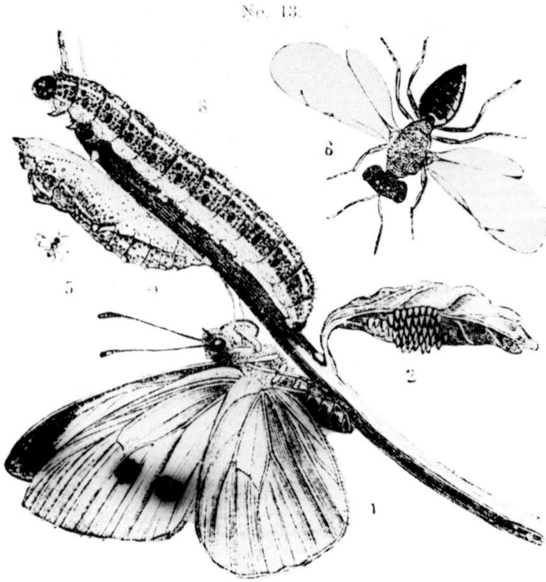

Left, Woodcut by by John Curtis. Fig. No. 13 from *Farm Insects*, *Pieris brassicae* L., the white cabbage butterfly. Adult, eggs, caterpillar, pupa, and parasitic ichneumon fly he called *Microgaster glomeratus* L *Below*, Woodcut by John Curtis Fig. No. 48 of *Farm Insects*, *Sitona lineata* L., the striped pea-weevil. This illustration shows one of the very few editorial errors in the work. The text refers to features numbered 3, 4 and 5 of this figure but these numbers do not exist; obviously they refer to another part of the woodcut which was cut when it came to printing the book

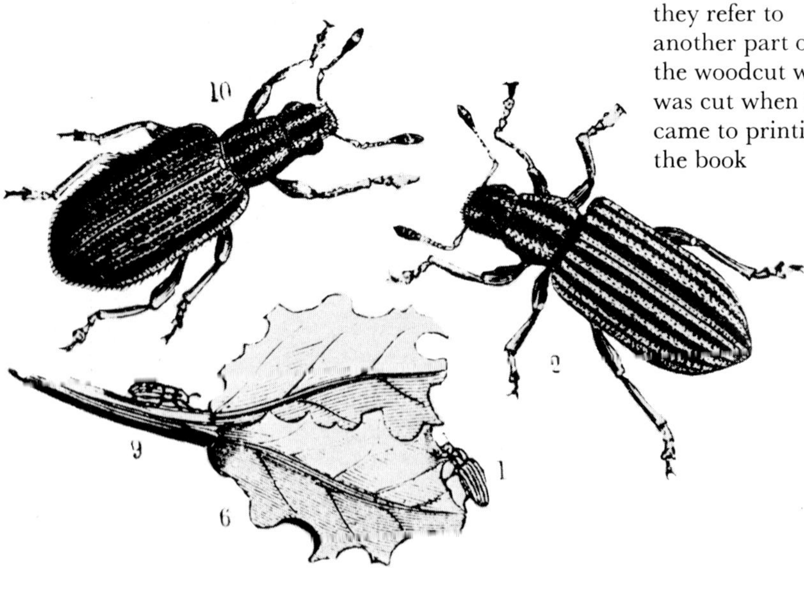

The ground lackey moth (*Clisiocampa castrensis*) drawn by the rival entomologists. John Curtis and J. F. Stephens were both producing serial works illustrating British insects, but though working at the same time only once did they feature the same insect, namely the ground lackey moth. Both are shown here

Right, the lithograph numbered Plate 13 in Stephens's book, drawing by C. M. Curtis (John Curtis's younger brother), with stone prepared by C. Wagstaff. The bottom moth is the male ground lackey *Clisiocampa castrensis*

Below, Plate 229 (1st Sept. 1828) of *British Entomology*, drawn and etched by John Curtis. It shows both male and female moths, the caterpillar, anatomical details, and the food plant. The John Curtis plate is so self-evidently superior to that of Stephens that it is difficult to understand why Curtis was so worried by his rival's publication

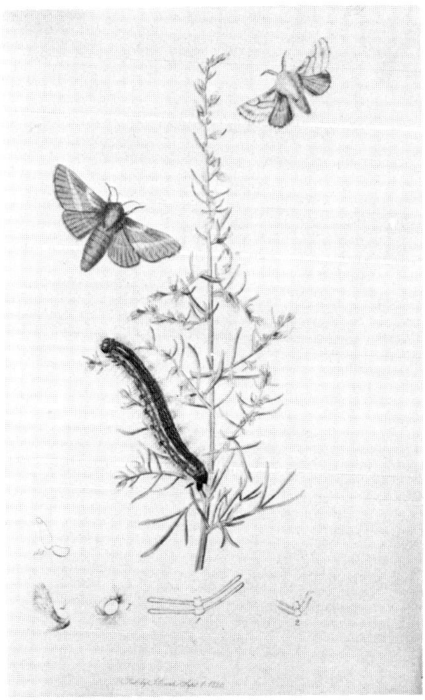

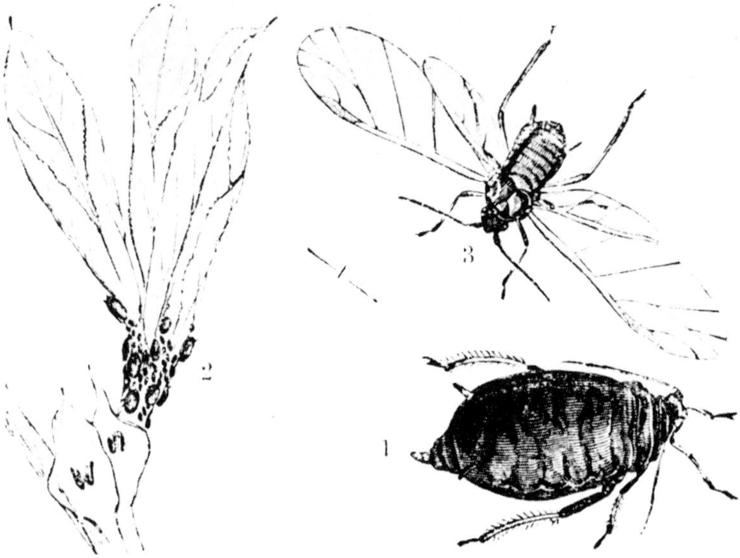

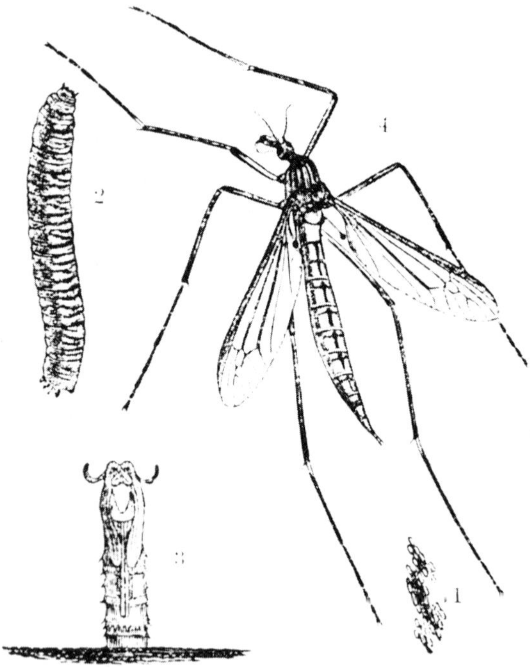

Above, Fig. No. 52 from *Farm Insects*, The black fly, *Aphis fabae*, on beans, a severe pest *Below*, Fig. No. 61 from *Farm Insects*, the crane fly (*Tipula oleracea*), known as Daddy or Old Father Long-legs, a pest of pasture and potatoes. 1. Eggs, 2. Grub, 3. Empty pupal case sticking out of grass, 4. Adult

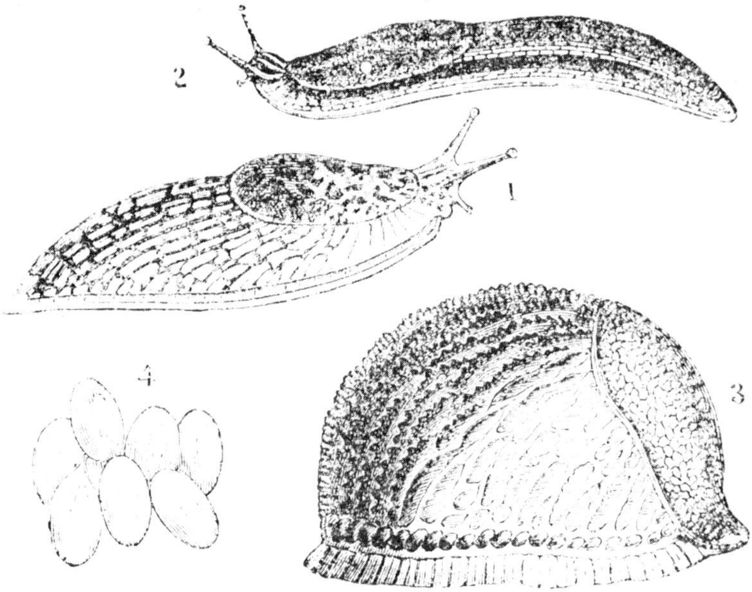

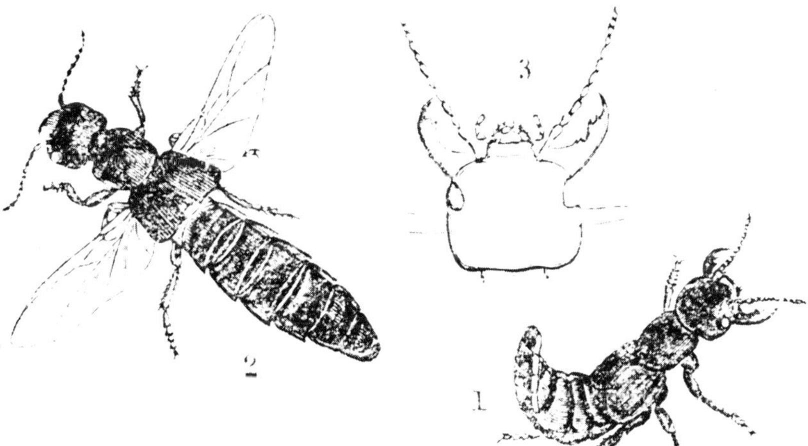

Top, Fig. No. 67 of *Farm Insects* included pests other than insects. Here is a fine woodcut of the common slug, *Limax agrestis*
Bottom, Fig. No. 68 from *Farm Insects*, the Devil's Coach-horse, *Staphylinus olens*; it is a hungry predator and destroys many pests

Plate No. 98 of *British Entomology*, The Narcissus Fly, *Merodon clavipes* Fab., often a severe pest of flower growers

Above, the top half of Plate M from *Farm Insects*; Some insects affecting the bean and mangel-wurzel;
1. Bean leaf eaten by *Otiorynchus picipes*, a. excrement of insect; 2. Female weevil; 3. Male walking; 4. Mandible or jaw of weevil; 6. Labrum or lower lip; 7, 8, 9, 10. Larvae; 11. A beneficial insect, *Cerceris arenaria*, a sand wasp which captures and destroys the weevil

Right, Woodcut by John Curtis. Fig. No· 50 from *Farm Insects*, Humble bees. No. 3 the Earth-bee, *Bombus terrestris*; and No. 4 the Wood-bee, *B. lucorum*. The former is shown boring a hole in a broad-bean flower the more easily to reach the nectar, thus reducing the chances of pollination and also damaging the set of seed in the pod itself

Miss Eleanor Ormerod, LL.D., Curtis's successor

7 Curtis the Man

In this chapter we shall endeavour to concentrate more on Curtis the man rather than Curtis the scientist, and this means going back a little. The information, though sparse, does suggest he was kind, likeable and seemingly with a sense of humour, as is suggested by some anecdotes in letters in the archives at Norwich[35] and those of the Royal Entomological Society: we quote extracts from a number of them as they give us an idea of Curtis's nature. His handwriting was small, fine, but generally very legible.

From C. W. Dale: Ladies usxd to like to teaze [sic] Curtis. Mrs Dale once gave him a specimen of the common housefly. He set it most beautifully. Then examined it and was very angry when he found what a common thing it was.

Obviously this was mock anger: one could not imagine that an experienced entomologist could set a housefly without noticing what it was.

The dedicated insect hunter is just that and nothing matters except insect hunting; lost appointments, food, drink, inconvenience to self and others are all naught. The above-quoted letter continues:

On another occasion Dale wanted Curtis to go with them to Round Chimney Pond. He would not go although Carpenter told him something good would be sure to turn up. And sure enough it did, in the shape of . . . [writing very indistinct, probably *Acentropus*, of which there is a figure in *British Entomology*]. He was very angry and said my father ought to have sent Miss Carpenter home to fetch him.

Alas! One does not know who Carpenter or Miss Carpenter were. How old was this female slave? Presumably not a child or she would not have been referred to as 'Miss'.

When Curtis came to London, as we have already noted, he met Leach and many famous men. Leach, apparently, was not always the dry as dust naturalist of the British Museum. Taking Dale and Curtis there on one occasion he jumped on the back of a giant scarab beetle and pretended to ride it; obviously the large basalt beetle of the Ptolemy period, circa 200 BC from Heliopolis, exhibit No. 74 of the Egyptian Collection, which is still there and has not been ridden again, one imagines, from that day to this.

Curtis married soon after he was established in London; it was probably about 1820 and really all we know about his wife is that her Christian name was Lydia, she was a Middlesex girl and had at least two children by Curtis, Edward, born 20 March 1821 and Elizabeth, born in 1826. Whether this name Elizabeth lends any grain of support (see Appendix 1) to the suggestion that Curtis's sister was Mrs Elizabeth Hay, aunt to the young girl, is a moot point. Elizabeth was a common name at that time.

In July 1825 Dale (senior) and Curtis went to Scotland, Curtis being the former's guest. They were abreast of the times for they went by steampacket; leaving London on 25 June they reached Edinburgh on the 27th, remarkably good steaming. They met interesting people and insects; among the former were Sir Patrick Walker (the entomologist) and Sir Walter Scott (the author), looking very splendid in the uniform of the Scottish Royal Bodyguard; of the insects they added thirty new species to the British list. They returned through England by land, walking much of the way with knapsacks on their backs.

In the footnote at the start of chapter 6 (page 41) we mentioned that Curtis eventually became a gentleman. This was a gradual process, small points being taken and consolidated along the way until the status undoubtedly was achieved. Rubbing shoulders with knights, magistrates, the wealthy, doctors and eventually peers led to the following entry in *The Royal Blue Book and Fashionable Directory for 1828:* 'Curtis, John, Esquire, 4 Grove Place, Lisson Grove, London.' Could one dispute that?

Curtis and Dale were great friends and Curtis frequently made trips with him and visited him at Glanville Wooton, Sherborne, Dorset. He went on a trip with Dale in Hampshire in July 1831, as he tells Haliday; he thinks it good for his health, but of course he loved it more because of the insects to be found and the 'communing with nature'.

Marriage did not at all seem to restrict Curtis's desire to travel and find insects, though the need for money did. There are frequent references to things such as 'the bad state of the Treasury' and the difficulty of being a scientist.

Conditions are not very cheering to anyone connected with the arts and sciences: the luxuries of life [including, one assumes, entomology] of course will be lopped off just before people allowanced themselves with the butcher and baker.

In 1830 Curtis was in France (going there via Jersey) and was most impressed with it, he writes:

I was delighted with my tour and astonished at the beauty and resources of the country. I think there can be no doubt that France is the most powerful Nation in Europe. . . . It is a land flowing with wine and oil if not milk and honey.

Did he consider Britain to be in Europe? Probably not. In France Curtis bought a collection of insects from Senegal, but paid so much for it that he was obliged to sell half. He complains also that collectors are only really interested in large insects.

In 1831 Dale carried Curtis off for ten days to the New Forest and the Isle of Portland. Although it was summer the weather was cold and bad and no rare insects were found; they used to be up and about in the forest by 4 a.m. Curtis this same year also visited Lord Malmesbury at Heron Court; it was late for botany or entomology but Curtis went to Stonehenge, which fascinated him. He found it more interesting than 'Old Sarum' and he was 'never more desirous of knowing the history of anything'. He thought that from a distance the stones might be a group of horses; the quantity of barrows and ancient roads showed that 'this desert was once thickly populated'. Obviously the place exercised as much fascination then as it does today but now part of the charm for us is the remoteness and isolation. When Curtis was there wildness was not greatly admired – the hang-over from the eighteenth century. Curtis was impressed because it 'was once thickly populated'.

In 1832 we find him regretting he cannot read German, and that what with the winter and the cholera in Germany he finds himself quite cut off from that country. The mails are delayed: he has not received letters from Welbeck Street – his sister's house (Elizabeth Hay?), obviously a more reliable address than far out suburbs like St Pancras. Another difficulty was entomological pins:

To Haliday, 2 March, 1832: The day I went into Suffolk I wrote under a frank . . . Dunsford's pins are excrable, but there are none so small anywhere else and Dale and I make the heads keep on by holding them a moment in the candle and touching them on grey sealing wax, which is well worth doing for rare insects. . . . Why do you not send me patterns? I will get them for you. I know it is useless sending [to Dunsford] as the man in the shop does not know a pin from a needle. (See Appendix 4 on dates.)

The 'franking' of letters was important to a man such as Curtis in those days: peers, certain officials and civil servants and members of parliament, were allowed free postage and all they had to do was to sign their names on the envelope. These favoured few could save their friends vast expense by 'franking' letters for them, even though the business covered by the franked envelope was nothing to do with the franker. All this ended with the introduction of Roland Hill's penny post. The civil service 'franks' its mail still, but only official letters. In fact today the only real crime among civil servants, compared with which espionage and breaking the Official Secrets Act are as nothing, is using a franked envelope for private purposes: the punishment is not generally known or specified, but is believed to be 'something with boiling oil in it'.

In October 1832 Curtis was in Ramsgate with Mr Walker. In December, though possibly still enchanted with France, he is less enthusiastic about the French:

23 December, 1832: I cannot say much for my correspondents in France. If they want anything they write, but with all their *politesse* I hear nothing of them till they again want something when they do not even acknowledge the receipt of some present. . . .

Curtis continues that he is going to describe the Straits of Magellan insects and he wishes Haliday a Happy Christmas and New Year, confident no doubt (note the date) it would be in Belfast before or on Christmas day (see Appendix 4). He is also upset by the death of an old friend, an eminent solicitor in his parish who was found drowned in a warm bath.

In February 1833 Curtis is working on the Magellan insects and wishes Haliday were with him so that together they might look over, 'Some of the wonders that delight me so much from tropical climates.'

One would hardly describe the Magellan Straits as tropical; perhaps some insects had also been collected on the way down. Curtis finds his dissatsifaction with the collecting world growing. This letter continues:

. . . but I find the richest naturalists the most stingy, I never can find anyone who will even go halves with me in a collection unless it is so cheap that I can take it all myself.

This is the year he was elected a fellow of the Oxford Natural History Society, an event which gave him much pleasure: it revived him from the state of bad health and depression in which he had been immersed, due to influenza, the death of a friend from it and the loss of a considerable sum of money from the non-delivery of a letter. In September 1833 he had dysentry and when he recovered and got back to his house (57 Upper Charlotte Street, London) he found scarlet fever there and so left to go to Heron Court again, presumably leaving poor Lydia to cope with the children and sickness. It seems rather heartless and unlike him. He says he would like to go to Ireland but bad health and an empty house prevent it. Perhaps, after the scarlet fever, Lydia had taken the children away to the country. Nevertheless he did spend a month in the autumn with Lord Malmesbury. In 1834 Curtis described the insects brought back by Captain Sir John Ross's expedition to the Arctic. The expedition had been attempting to find the north-west passage, had been iced-in for years and at home the members had been given up for dead. Curtis also etched and hand-coloured a very fine plate of fifteen insects, dated 1 May 1834.[67]

In a letter to Haliday of 21 May 1834 Curtis is planning a visit to Scotland with Haliday and this takes place in July. They visited Skye, Arran and the falls of the Clyde:

all of which bring back to my recollection some of those hours which make me attached to this beautiful World in spite of its inhabitants.

Curtis takes a berth on the *James Watt*, leaving London on a Saturday night and expects to be in Edinburgh on the Tuesday at 10 a.m. when he may go straight on to Glasgow, which he did not like:

the only thing that may induce me to stop at the modern Athens is the fine scenery and the detestation I have of Glasgow which is the filthiest place I have ever seen and even Mrs (or Mr?) Trollope would think favourably of the Americans after a visit to Klasscoe as they call it and the Inhabitants so disgustingly vulgar.

We do not know what the trip was like, but the journey home was pretty terrible according to Curtis, who says he is glad to have heard from Haliday that he is safely back in Belfast:

28 October, 1834: dreadful journey . . . weather turned out very bad and we had to change the boat at a Loch and go by an Omnibus a mile, my luggage being all the time exposed to heavy rain, which with running foul of a Vessel and occasionally of the piles from the Helmsman being asleep (a very pleasant thing going 10 miles per hour) and not reaching the station until long after dark, made me determined never to go that way again.

This, bad enough, was followed by a dramatic drive through a tunnel. The station referred to must have been one of the new-fangled railway stations. Curtis was always a pioneer and trying new methods both in science and life. Curtis finally reached some Assembly Rooms, which one presumes he was making for, in the pouring rain; his umbrella was taken and the attendant said, 'What can you expect on a night like this?' Perhaps it was in his detested Glasgow. On the notice board was:

If the gentleman who accidentally abstracted an umbrella from the Assembly Rooms will call at the *Police Office* he will there find the *Case* which is no longer any use to the owner.

It was a big meeting: 1,400 members were present, including 500 ladies.

. . . everything went off so well and I enjoyed myself so much that I very much regretted you were not with us.

In 1834 Curtis was concerned with a case in Chancery: it was the time when delays there were infamous and led Dickens to castigate them in 'Jarndyce v. Jarndyce'.

4 July, 1834: Unfortunately some property of importance to me is in Chancery but I understand nothing will happen until the Autumn and I trust a man who owes me a large sum will return to England in a few days.

Seven months later, needless to say, the business was not settled:

22 February, 1835: . . . and what is even worse one of the trustees who had a considerable sum of money in his hands is gone all to pieces. I am certainly very unfortunate in money matters and I am very melancholy in consequence of the dangerous state of health of one of my kindest friends and relatives, whose death would be a great affliction and cause me much trouble of which I have enough already.

One feels this person was his possible brother-in-law, Mr John Hay, of 75 Welbeck Street. If so he recovered for he was alive, aged seventy-five at the 1841 census.

Curtis was now well-known as an authority on insects; collectors valued his opinions on identification and naming in spite of what his rivals might say and people flocked to his house in Upper Charlotte Street. He was so besieged that he had to make an announcement that he could only receive his friends on Wednesdays between the hours of 11 a.m. and 4 p.m. 'Friends' can be taken to mean his customers and acquaintances wanting free advice. Do the hours suggested throw any light on meal times? Curtis was not a fashionable man. He would be up early, breakfast about nine and have his dinner at 4.30 p.m., having got rid of his customers; perhaps a real friend or two might be invited to stay. A supper would have been taken at 9 or 10 o'clock.

This same year (1835) sees Curtis off on a tour of Ireland and he differs with his companion, Captain Heigham (writing indistinct and thus the spelling doubtful) over the route. The gallant captain wants to see the caverns of Mitchelstown whereas Curtis prefers mountain views. He would rather travel a thousand miles for such a view than to cross the Thames to burrow into a cavern. The journey was difficult as he describes in a letter dated 30 June 1835 from the Gloucester Hotel, Clifton, near Bristol. Curtis missed the boat there and had to take another steamer – the *Killarney* – which after setting off had engine trouble. There was a tremendous hammering as the engineers were cutting new cogs on board. Finally they were towed back to port with the pumps working the whole time. It would appear that they had had a narrow escape from foundering. It was an age of enterprise as the brave attempt to cut new cogs on board shows. Curtis eventually got a passage to Cork on the *Victory*.

There is no account of quite where they went in Ireland. He seems to have met his new friend, the Earl of Malmesbury, there and to have been taken ill – 'Rheumatism, slow fever and the "Galway Mulligrubs".' He returned with an easy passage to Holyhead, slept at Bangor and walked to see the Menai Bridge:

6 October, 1835: . . . which was one of the *few* things that exceeded my expectations, and whilst contemplating it I could not help feeling astonished at the power of the human mind. It must have been a glorious day to Telford when the last stroke was given to that splendid work.

The following month his health is better; Dr Farran has cured him of the Galway Mulligrubs, but whether the cure was 'goats' milk from a Kerry cow' (which sounds a contradiction) or the cause of the trouble is difficult to devise. He appears to be much in with the various Harrises – Charles, the Bishop of Winchester and Lord Malmesbury, and they discuss the insects, politics and religion of both Ireland and Iceland. He was probably not much interested in the latter. We have already seen he did not think much of caverns and in the following year we find him expressing similar views about the arctic north. Curtis had seen Walker, the entomologist, and says:

2 June, 1836: perhaps he may not tell you that Christy, the botanist, a son of Sir William Hooker and himself are going to the North Cape, what for I can only imagine, but not for the purpose I suppose that first suggests itself to you, and certainly if I had time and money I should not spend it in freezing myself. I would at all times rather *pay* for ice than get it for *nothing* first hand.

This letter raises two interesting points: firstly what was the object of the expedition if it was not that which would first suggest itself, namely the collection of rare insects and plants to add to scientific knowledge and the richness of gentlemen's cabinets? The mind boggles: espionage, homosexuality, sport, the midnight sun? The second interesting point is that ice was an article of commerce. The rich always had their ice-houses, filled with blocks during the winter and used during the summer, but apparently it could be purchased too. London supplies tended to come from the Thames marshes where farmers collected it as a crop and sent it to a large ice-store at Barking. It was also imported, packed in straw from Scandinavia, and even from America (Wenham ice from Boston), which is surprising in view of the length of the voyage. Most of it was used for the preservation of fish. In this same letter Curtis says he is on the eve of visiting Yorkshire and is much occupied in endeavouring to increase his income 'in which my friends have again failed and consequently I am not in very good spirits'. This is most likely to mean that various friends owed him money for insects supplied to them.

In June 1836 he is finalizing the second edition of his *Guide* and in November is deciding to devote a year to botany, a thing he was never able to do although he had a good knowledge of plants.

In May the following year (1837) Curtis is so hard up that he has to move from No. 57 Charlotte Street[63] to No. 11 Robert Street,[64] Hampstead Road. The curious thing is that the rent of 57 Charlotte Street is entered in the rating book as £60 per year in 1837 whereas the rent for 11 Robert Street is slightly more – £62 per year, but the tenant at Charlotte Street is given as Noel N. Carter, who possibly was rack-renting it to the unfortunate entomologist. At Robert Street Curtis is entered as the tenant, renting it from George Baker: the rates were less, £2. 19s. 7d. against £8. 16s. 3d. for the former house, so that Robert Street was a little cheaper than Charlotte, even if the latter were not sub-let. The house in Charlotte Street still stands and is larger and much pleasanter than the Robert Street one, even though the latter had small balconies to its two first floor windows. It has recently been pulled down for a large housing development scheme.

About this time Curtis noticed a severe pain in his left side, probably the start of the trouble that eventually killed him.

Only one certain letter has survived for 1839, the year his brother, Charles Morgan, died: he used black-edged paper for it and in the early part of the following year too. His financial troubles were by no means ended.

There is, however, an undated letter which may belong to 1839 in which Curtis bemoans his financial state and the volume of work devolving on him to keep *British Entomology* going:

> Drawing, engraving, finding a considerable sum of money, more than I receive, it is too much for one person to accomplish.

In this letter Curtis has some religious reflections. It staggers him to hear that Paley and Abbott write dangerous books, the former's presumably being *A View of the Evidences of Christianity* (1802) and Haliday contending that one should just believe and not seek to prove one's belief, a proposition much at variance with Curtis's scientific soul. Abbott was probably Charles Abbott, 1st Baron Colchester (1757–1829), a scholar and at one time speaker in the House of Commons when that task was more arduous than it is today. Abbott strongly opposed the admission of Roman Catholics to parliament. Curtis says the only recourse for poor wights like him is to go back to the Old and New Testaments. He is still toying with the idea of leaving England.

I am surprised to hear from a friend of mine that £80 per annum would well maintain an *English Gentleman* at Nice and enable him to keep a horse.

We find a chatty letter to Haliday in February 1840 (Appendix 5). He is sorry to hear that Haliday's eyes are weak and in his own case he found working with glasses by candlelight caused his eyes to suffer so he gave it up. Now he only reads or writes by candlelight and never draws. His remarks here lead us to another facet of his life. He used to visit the theatre:

11 February, 1840: (on candles): . . . which how [*sic* however] wretched light they nowadays give do not inflame the eyes like a lamp. You may smile at my philosophy and I do all in my power to get a better light with candles, but after all a wax candle is sufficient to write by and it seems to me that the heat as well as the light from a lamp is pernicious. I never go to a theatre without being almost blind the following day.*

Ah, things were not like the good old days, even candles! Curtis says he has made sacrifices in order to continue his great scientific work which has obliged him completely to reverse all his hopes and plans. He cannot retire (he was forty-nine years old) but must start life *de novo* and is looking for an official appointment in order to increase his income and hopes his influential friends will help.

Indeed, I have had quite enough of Natural Science as a Profession and should greatly prefer making it in future an amusement.

Nevertheless, whatever post he was thinking of, he was caught in the trap and soon describes how he is toying with the idea of an *Atlas of Insects*, to describe insects in a concise natural way for the public, but he

will never be betrayed into the folly of going to press until I am secured from loss by an ample list of subscribers.

He is having no more to do with periodicals and is fighting against time. Here we see the start of the idea for his important

* One of the reasons Victorian rooms were not cold on winter nights was that the fire was much supplemented by gas lighting and candles.

life work *Farm Insects* about which we shall have much to say below.

February steeped him in gloom: his Robert Street house faced north; one presumes he took the first floor room as his office and study, even so it was so cold near the window that he could not dissect or draw, only read and write. Lydia had to do without a drawing room, but did have a maid. However 'ruined' one was, one notices in the Dickens and Thackeray novels, one always had a maid.

> . . . bookbinding, advertising and Reviews have given me no little trouble. I am afraid of a wet summer. I heard of an Italian who said he would take any wager we should have 200 days of wet weather . . . Entomology seems to be paralysed, as well as everything else here, our circulasion [*sic*] is very tardy. If all the Opium destroyed in China had been consumed at home we could not be in a more drowsy state, but I suppose the Chartists will waken us bye and bye.

Curtis records a pleasant evening at a dinner party given by Sir Henry Willocks at which he met Major Davies, who was going to Barbados, Dr Robinson of Armagh and Sir James South,* knighted by William IV in 1830 to prevent him, after some learned quarrel, retiring in dudgeon to France with his telescopes. Coulton had become ill from travelling outside on a stage coach and the party hoped he would recover his health in South America and give his work in Mexico to the world. It is curious that Curtis apparently considered Mexico to be in South America; it is, of course, in North America and just before that time the country extended still further north, Texas then being part of Mexico, it not yet having been conquered by the USA.

In April 1840 Curtis still had hopes of an official appointment; something of the nature of the post Stephens had had, which would enable him to study entomology as a science rather than as an entomophile's serving man. At this time J. E. Gray, who, though a botanist, had been engaged by J. E. Children to catalogue the reptiles in the possession of the British Museum and who had been the first secretary of the Entomological Society, was given Children's post on the latter's retirement – the Keeper of the Zoological Department of the British Museum – a position Curtis had wanted. Resigned however, Curtis hoped to obtain Gray's former position and much lobbying took place.

* The astronomer.

The patronage system was still strong. By May 1840 Curtis knew he had not got it and was very disappointed, but he was aware it was not Gray's fault and remained on friendly terms with this remarkable man. Curtis was in sympathy with many of Gray's ideas, both zoological and social, many of the latter being advanced for the age – public playgrounds for children, sanitary reforms to stop cholera, provincial museums and free entry to all museums (would he were alive today), and uniform postage, prepaid by means of stamps. He was considered to be a 'dangerous radical'.

The 1840 letters clearly show the doubt, hopes and difficulties of this period of Curtis's life. One of 13 March, to Haliday, starts off in an ordinary way; he has sent insects from Germany and France worth 26s. and goes on to say he has been much occupied in writing to 'Deserters', that is to subscribers and institutions who have discontinued their payments for *British Entomology*.

To the credit of the former my applications have been tolerably successful, but *I dare not tell you how disappointed I am at the results attending the latter*. I fear there is not a chance of my ever recovering my Property from the unfortunate dilemma in which it has been placed by my own confidence. I am convinced Entomology is on the Decline, which I am truly sorry for, not only on my own account, for I am too . . . an . . . of the works of Nature to be influenced by fashion, but I regret the want of better feelings in the Higher Grades of Society who alone ought to cherish such invaluable and delightful pursuits.

If by 'Entomology' Curtis meant the collecting mania it may well have started to lose its intensity at the dictates of fashion at that time.

In April 1840 Children sold a considerable number of books and Curtis purchased some for Haliday and some for himself. Among others were three volumes of Audouin costing 21s. for Haliday and six volumes of Dejean, 36s., and a Fabricius at 48s. for himself. Books of this nature were expensive luxuries: £2. 8s. is a lot of money when it is considered that a gentleman could live comfortably (at Nice) on £80 a year and keep a horse. The sale, obviously an auction, caused him 'a sad loss of time'. He probably helped Children with it, both for friendship's sake and with a view to getting his post. As noted above, the blow fell in May.

9 May, 1840: [Curtis has been much occupied with Museum affairs] . . . such a state of anxiety, and after all I can get nothing. It is thus that men engaged in Natural Science are rewarded in this country and I am so thoroughly disgusted that if I get nothing this year I have fully made up my mind to sell off everything and go abroad in the Spring.
12 December, 1840: I think Nature as well as Man is become tired of Enty. for I could find nothing new or worth having [he is planning to go to the country] . . . I am thoroughly disgusted at the results of my labours and devotion to Enty. and I hope I may then enjoy the pleasure to be derived from the study of Nature free from the petty fogging and beggarly crew that make a stepping stone of us here in London.
Through my extreme folly my income is so limited that I dare not be idle whilst I remain here and I am therefore collecting information relationed [*sic*] to the Insects injurious to the Agriculturalist, Gardener, Planter, etc.

When this great idea struck him he was obviously turning over various schemes in his mind such as living in Italy (at Nice, then in Italy) or in the English countryside. He needed an extra £100 a year (later reduced to £50) to do it and fortunately for the world, if not for himself, he did not get it and turned back to the injurious insects proposal.

Curtis did not join the Entomological Society until 1851; a reason may have been its too close connection with the world of collectors.

The lack of interest in economic entomology shown by its members in its early days may be seen by the fate of the *Prize Essays* offered by the Society. The first offer was made in June 1834, the subject being the turnip fly (see page 83), an important crop pest. A year later no composition had been submitted and the period was extended for another year. One entry was received and the prize – a handsomely bound copy of the Society's *Transactions* – was awarded to Mr H. Le Keux, subsequently the engraver of many plates for Curtis's *Farm Insects*. The next offer was for an essay on the pineapple scale, for which there was no entry. The third was the turnip sawfly: three entries were received and a prize was awarded to George Newport. And so the matter continued. No entries were received in 1839 on *Agrotis segetum* of turnips, nor on the pests of apple and pear trees, nor on the pests of market gardens. However an entry was received in January 1844 for a *Bibliography of Entomology* and the prize given to a Mr Evans: the subject, it will be noted, was more 'pure entomology' and had little to do with crop pests. One wonders why Curtis did not submit entries for at least some of these prizes. At any rate it would seem that Mr Le Keux must have drawn largely on Curtis for his material.

The but desultory interest in the prizes continued:

1844	Natural history of *Stylops*	No entry
	Natural history and Medical Properties of 'Vesicatory Insects'	,,
1847	12 Species of Tortrices	Apparently no entries
1852	Noxions or beneficial insects changed to:	No entry
	The duration of life of three kinds of honey bee	3 entries
1853	General natural history of the species of the genus *Coccus* injurious to fruit trees	No entry
1855	Lac dye of commerce	,,

Perhaps the prizes offered were not tempting enough: they were five guineas or various splendid books usually especially bound.

In other quarters it was at last beginning to be realized that pest attacks on crops were not Acts of God that had to be endured but events which Science, with a capital S, might control. If it could bridge the Menai Straits and reduce cholera attacks, surely might it not also prevent the murrians and swarms of caterpillars and beetles attacking man's food, causing such devastating losses? The idea appealed to the Royal Agricultural Society, after all its motto was 'Science with Progress', and the first requirement for such an advance was knowledge. The news, and the actual examination, of Curtis's collection of 'injurious insects' clinched the matter and the RAS commissioned him to write on crop pests. The scheme had the support of the government and particularly that of Sir Robert Peel who was anxious to increase farm output.

His first subject was the turnip crop, one of much greater importance then than now (2½ million acres then against 350,000 now) and in order to do this he visited Norfolk and Suffolk to discuss matters with farmers, to find out what they knew, wanted and needed to know. Curtis thus got the feel of the situation. The series continued until 1857 and we discuss it in the next chapter.

The articles were a great success and as a result Curtis was commissioned by the great Lindley to write on pests for *The Gardeners' Chronicle and Agricultural Gazette* which he did, using the name 'Ruricola'.

We are now in 1841; from then until 1846 the main source of personal news about John Curtis – the letters to Haliday – dries up. In 1843 he moved to Hayes, Middlesex, to enjoy the country air. He also made a tour of Italy. He was enabled to do this because he had at last got the appointment he needed, for which event we must go back a little.

Curtis was sending Sir Robert Peel, the prime minister, regular reports on insects injurious to the farmer: for instance, the fifth went to him in October 1841. In 1842 Lord Ripon requested Peel to put Curtis on the pensions list. On 4 November 1842 Sir Robert wrote to Curtis advising him that he had been granted a pension of £100 per annum so that he could continue to work on insects injurious to vegetable crops that feed man or cattle and find 'the best mode of preventing or repairing their ravages'. This pleased both Curtis and Lord Ripon, but such was the vanity of the latter that he wrote to Peel saying that whilst there was no doubt that Curtis was deserving he felt that Peel had given Curtis a pension as a personal favour to himself. To his Lordship a personal favour was more likely to be the cause of such an action, that is patronage, than the merits of an individual or the possibility of saving thousands of tons of food from destruction!

Lord Ripon was born Frederick John Robinson, the second son of the second Lord Grantham, and became an important politician, in fact prime minister, for a short while after the death of Canning in 1827. Previous to this, 1823, he had been chancellor of the exchequer when his cheerful speeches during a time of trade depression earned him the name of 'Prosperity Robinson'. In 1833 he became the Earl of Ripon and in Peel's second government (1841–6) was successively president of the

Board of Trade and the Board of Control, hence his importance to Curtis.

Curtis, naturally, wrote to thank Sir Robert Peel; he says that he derives much satisfaction from his work and now with his pension he can devote time to investigate

> insects injurious to man and I hope to benefit the Cultivators of the soil and unfold the hidden leaves of Nature's book for the Agriculturalist and Gardener so that they may become more like men of Science and improve their instruction.

The paragraph has a splendid Baconian ring. One might deduce that Curtis knew the *Novum Organum* and the need to 'uncover the secrets of nature'. Scientists stood in high popular estimation then; they are tending to lose that position today, as witness a large graffito on a wall by the Moorfields Hospital, rapidly fading though still there: 'Down with so-called scientists', a cry from some poor soul's heart.

We know very little of Curtis's life at Uxbridge, except that he did not like it there. His original idea had been to retire to a simple life of slippered ease, a recluse aged fifty-two, in the depths of the country and now he found himself with his new career of economic entomologist busier than ever. He had to travel a great deal to ensure accuracy in his new work and Uxbridge, although on the Paddington canal, was six miles from West Drayton station. There is still an inn on this canal at Uxbridge called the 'Paddington Paquet'. He returned to London in March 1846 on the grounds that the soil and air at Uxbridge did not suit him. Travel was becoming much easier; by 1845 there was a main line to Curtis's beloved Norwich. One could reach Newcastle and Carlisle in the north and the lines were pushing on to Edinburgh and Glasgow. In the south the railway went to Dover, via Reigate, Tonbridge and Ashford; it got to Exeter via Reading and Bristol and work was in progress to take it to Plymouth; a whole new world of cheap and rapid transport was being created and the 'Parliamentary Train' (1845)*, at first much resented by the railway companies, actually made their fortune.

* The 'Parliamentary Train' was an obligation on all companies to run at least one train each way every day at a fare of 1d. per mile, stopping at all stations: it opened the railways to the common man and travel ceased to be a privilege fo the well-to-do.

In 1846 Curtis resumed his correspondence with Haliday. Curtis tells of his visit to Italy, how he enjoyed it, travelling by the coast road to Genoa he eventually got to Naples. He saw many of the pictures and palaces he had hankered after for years. He returned via the Tyrol and the Rhine. Obviously his pension and new profession had improved his finances. He was living at 'Hayes Court', which name does not suggest a cottage. Perhaps he felt a little ashamed with Haliday at abandoning pure science and has to explain it:

11 May, 1846: Finding that pure science led, like Poetry in the olden times, to starvation, I turned my attention to economic Science and engaged to write Reports for the Royal Agricultural Society on the Insects injurious to the Farmer, ten of which have been published.

Curtis also mentions he has been doing the insect work for the *Gardeners' Chronicle* since its commencement so he finds himself just as busy in the country as in London. He goes on to say that the Hayes air does not agree with him, that there is a clay soil and that he feels at least ten years older than when he was in London. Once more he is thinking of going to live abroad, but it can only have been a passing thought because his new job was rewarding both in terms of money and, more important, in itself; he was supplying a much needed aid to farmers. He could not have continued this work had he gone abroad.

The month before (April 1846) he sent his tenth report to Sir Robert Peel and tried to get more money from the Civil List, as this is the last report he 'will be permitted to write for the Royal Agricultural Society'. He explains that when he got his pension he retired from London in order to work more cheaply but finds he cannot do it on such a small sum and he needs the London contacts. Please could he have a pension on the same scale as Mrs Brown and Mr Owen? Curtis points out he spent sixteen years and £12,000 on the publication of *British Entomology*: he has 800 drawings, valued at £15,000, which he wants to deposit at the British Museum (they may be seen there to this day) and he only wants to continue working for the benefit of Agriculture. He does not know Sir Robert personally but Lord Malmesbury and Lord Ripon know him and for many years he has been engaged with Lindley in literary pursuits of benefit to the Farmer and Gardener. Mrs Brown was probably the widow of an entomologist, a member of the Entomological Society in

1838, who died in 1840, about whom little is known, not even his initials.

Richard Owen (1804–92) was the son of a rich Liverpool merchant trading with the West Indies. He became a distinguished zoologist and superintendent of the Natural History Section of the British Museum. In 1838 he founded the Royal Microscopical Society and was its first president. In 1842 he accepted a Civil List pension of £200 from Sir Robert Peel but refused a knighthood, probably on the grounds of the expense it would incur, for in 1884 he took a KCB from Gladstone plus an extra £100 of pension. Probably Mrs Brown was getting the same £200. In an age of great reticence everyone seemed to know what these pensions were. One can see it would be very galling to Curtis to think he was but half as valuable as Mrs Brown and Richard Owen. Being the son of sign painter, rather than of a rich slave-trader, may have had something to do with it. Perhaps, like Mr Salteena, he was not quite a gentleman.

It is interesting to note that Alfred Tennyson in the early 1840s sold his small Lincolnshire estate and invested the money in the 'Patent Decorative Carving Company' (see E. Burton[76]). The company failed and left the poet ill and penniless. Having the right friends and connections, they applied to Peel, who granted him a pension of £200 a year. £200 seemed to be the going rate for gentlemen, or even ladies, in what was known as 'reduced circumstances'.

One wonders why Curtis says the tenth report was the last one he would be permitted to write. It is probable that this request for an increase in pension was his method of drawing Peel's attention to the fact that rivals were nibbling at his job within the usual competitive spirit of the age. Sir Robert seems to have intervened on two grounds at this point, firstly to keep Curtis at the work at which he was so good and secondly to save the government money, for the tenth was not the last report Curtis wrote and his pension was not increased at this time. One must remember that by now governments throughout Europe had had a considerable scare as to what pests could do, for example in the potato blight incident. Though not caused by an insect there were those who held that it was. A learned Mr Alfred Smee, surgeon to the Bank of England, proved in 1846[71] that it was due to an aphid, the fungus observed by the Reverend M. J. Berkeley being the *effect* of the aphid attack, not the cause of the blight.

The blight had starved thousands and thousands of people, particularly in Ireland. Scientists of Curtis's calibre, therefore, were valuable.

In addition to his economic work Curtis continued his contributions to the scientific journals and learned societies of the age.

In the autumn and winter of 1850 Curtis was well enough off to visit the continent again. He went to Nice, Genoa, Turin and the north of Italy, coming back through the Tyrol and Switzerland. We have news of him in the 1851 census when he was living at 18 Belitha Villas, Barnesbury Park with his son Edward, a clerk at the Bank of England, his daughter-in-law Emma and a grandson Herbert Edward Curtis, aged seven months. There is no mention of daughter Elizabeth, now aged twenty-five; perhaps she was married and had gone away. In this year Curtis became a member of the Entomological Society, which was surprisingly late, in view of his lifelong interest in the subject. Once more (the winter of 1851) he went abroad for the winter, this time to Pau and a number of towns on the shores of the Mediterranean, and Venice, Florence, Lombardy, Switzerland and France.

This year saw him start another popular work when he contributed the articles on pests in *Morton's Cyclopedia of Agriculture* (1851–5). The series was illustrated by numerous woodcuts from blocks he had made himself.

Curtis's interest in putting entomology to practical use for farmers did not mean that he neglected scientific observation. For instance, dealing in the *Gardeners' Chronicle* with the nut weevil (his name for it was *Curculio nucum*), an insect which could cause considerable damage, he noted that two larvae were never found in the same nut. He concluded that the female weevil knows by some means or other when an egg has been laid on a nut and thus avoids it. We now know that the female, after laying an egg, 'marks' the nut with a special odiferous substance which inhibits egg-laying by other females subsequently visiting the nut.

Many scientists and learned societies at home and abroad appreciated Curtis's work and honoured him. The great Baron Cuvier described his *British Entomology* as 'the paragon of perfection'. In 1831 Curtis was elected Corresponding Member of the Royal Georgofili Society of Florence. In 1831 he became an honorary member of the Ashmolean Society of Oxford and in 1836 an honorary member of the Academy of Natural Sciences of Philadelphia. He was a member of the Entomological Society of Paris from 1834 and was made an honorary member of that body in 1855 when it was learned how poor he was. In 1849 he was elected Corresponding member of the Natural History Society of Nuremburg. He was welcome wherever he went all over Europe and no doubt would have been welcomed in America had he ever reached there.

FARM INSECTS:

BEING THE

NATURAL HISTORY AND ECONOMY OF THE

INSECTS INJURIOUS TO THE FIELD CROPS

OF

GREAT BRITAIN AND IRELAND,

AND ALSO THOSE WHICH

INFEST BARNS AND GRANARIES.

WITH SUGGESTIONS FOR THEIR DESTRUCTION.

By JOHN CURTIS, F.L.S.,

HONORARY MEMBER OF THE ASHMOLEAN SOCIETY OF OXFORD, AND OF THE ENTOMOLOGICAL SOCIETY OF FRANCE; CORRESPONDING MEMBER OF THE IMPERIAL AND ROYAL GEORGOFILI SOCIETY OF FLORENCE, OF THE ACADEMY OF NATURAL SCIENCES OF PHILADELPHIA, ETC.

ILLUSTRATED WITH NUMEROUS ENGRAVINGS.

BLACKIE AND SON:

FREDERICK STREET, GLASGOW; SOUTH COLLEGE STREET, EDINBURGH; AND WARWICK SQUARE, LONDON.

MDCCCLX.

Facsimile title page of Curtis's *Farm Insects*

8 'Farm Insects'

The splendidly Victorian title of John Curtis's great work is set out on p. 77, using it will be noted, ten type faces, the epitome of Victorian grandeur. We deal with it here because the work is essentially a reprint with occasional additions of the articles from the *Journal of the Royal Agricultural Society* (1841–57).

Farm Insects was published in eight consecutive monthly parts at 3s. 6d. each or 22s. the set, a saving of 6s. Two thousand copies of parts 1–7 were printed and only 1,500 of part 8. The reason for this drop is difficult to understand as the whole series had been an immense success from the start. A possible reason is that the last section dealt with potatoes, clover and pasture, and potatoes in those days were not considered to be an agricultural crop. Even as late as 1879 – 82 when a Royal Commission on Agriculture was investigating the terrible depression in farming, the commissioners reported that they would not take much notice of the evidence from a certain Lincolnshire grower that he was prosperous, because he grew 200 acres of potatoes and thus was not a farmer but a horticulturist. To the Commission, farmers grew corn and that trade was depressed, due to American imports, which same imports greatly benefited stock raising, particularly in the north-west of England.

The publication resulted from the tremendous demand for the *Journal of the Royal Agricultural Society* containing the Curtis articles on pests. They ran from 1841 to 1857. It is an important book because it was the first to be written for the farmer by a scientist who knew his subject, who was accurate and who could present his material well. Moreover it was comparatively cheap and had most accurate coloured plates, making identification of friend and foe easy. Another unusual feature of the work was the arrangement of the material, which was by crops rather than by the different families of insects. This enabled the farmer to read the work with interest and relate control measures to practical ends. Of course there had been other works on pests and their control, but they were mostly on a single pest or crop or were vague, unscientific and unillustrated. An example of the former is Duhamel de Monceau and Tillet's magnificent book on the Angoumois grain moth (*Sitotroga cerealella*) of 1762[30] and two examples of the latter are J. Major's 1829 book on fruit tree pests[48] and Thaddeus William Harris's *Insects of Massachusetts Injurious to Vegetation* (1841).[34] Major's book has only a vague rambling description of the insects and no pictures. The American book is a scientific study arranged according to the orders of insects and very much reflects the influence of the collector rather than that of the farmer, although it was ordered by the State Legislature. It had no pictures. Curtis was the pioneer with *Farm Insects*. Its success and utility may well have influenced the American authorities for, in 1862, Harris's study was reissued under the editorship of Charles E. Flint. The second edition of that work, though very well illustrated and much more practical, is still arranged according to insect orders making the task of a farmer trying to track down a pest somewhat difficult. The American Curtis turned out eventually to be C. V. Riley (1843–95) – an immigrant from England who became the first Federal entomologist of the USA.

Farm Insects on the other hand is a thoroughly useful book and set the style for the many others that continue to this very day. Curtis pretty well invented the subject of 'economic entomology' and the changed attitude he imposed may be seen in his introduction. He starts off by saying he is not going to write an essay on agriculture, which he considers to be the basis of the wealth and prosperity of the country. He is thus opposing the Manchester school, who believed that Britain should be 'the workshop of the world' and import cheap food. He continues:

> Whilst Chemistry and Geology have lent their valuable assistance in rendering the land more fertile, little attention has, comparatively, been paid to those noxious animals which annually consume an amount of produce that sets calculation at defiance; and, indeed, if an approximation could be made to the quantity thus destroyed, the world would remain sceptical of the result obtained, considering it too marvellous to be received as truth. (p. ix.)

The amount of damage pests cause is still a matter of some debate and, in fact, is under investigation by the World Food and Agriculture Organization. It is largely a matter of how one measures losses, in tons of produce destroyed, or in money, or in land used to feed pests (see below p. 89). The two are not necessarily consistently proportional to each other (see Appendix 5) and the confusion produced in assuming that they are so, still leads to results too marvellous to be received as truth. The fact is that in a largely ill-fed world, pests consume a vast amount of produce that can be ill afforded and Curtis knew this. On the other hand, farmers as a group make more money from a short crop than an abundant one, so that if they overcome a common pest, the drop in prices may lead to a lower income for all of them. Curtis was also aware of the need for accuracy and experiment; he starts chapter 1:

Unless we collect facts on good authority, and conduct experiments with care and perseverance, our labour will be lost, in studying the economy of the insect tribes; for in the investigation of such living atoms, as they often are, the slightest error may lead us far from the truth. (p. 15.)

Another innovation in *Farm Insects* is the summary at the end of each chapter, very reminiscent of the modern textbook. The book is remarkably well illustrated; it has sixteen steel engravings, hand-coloured, the drawings by Curtis and the engraving by J. H. Le Keux. There are also sixty-nine woodcuts, probably by Curtis himself. All are exceptionally clear and well done and the story related of Curtis complaining to the engraver dates from this period. When pulls from a plate were brought to him by a Mr Smith the devoted entomologist said, 'But I drew thirteen lines there and you have only put twelve.' Curtis was among friends at the time and they all laughed, no doubt a most fortunate thing for Mr Le Keux.

There are 524 pages of text, distributed among the crops as shown on page 81.

Much light is thrown on Curtis himself and on the book by an examination of his introduction, from which we have already quoted. In a continuation of the above-mentioned passage he deplores the acceptance attitude of farmers to the ravages of pests but ignores the social implications:

If the same diligence were exercised by the farmer in destroying rats, mice, sparrows, etc., as is shown for the preservation of game, the producer and consumer would alike be benefited to a vast amount. (p. ix.)

Pagination of *Farm Insects*

Chapters	*Subject*	*No. of Pages*	*Percentage*
Introduction	Introduction	14	3
1–5	Turnips	137	26
6 and 7	Wireworms in Turnips and Corn	59	11
8–11	Corn	131	25
12 and 13	Peas and Beans	60	11
14	Carrots and Parsnips	24	5
15	Potatoes	48	9
16	Clover and Pasture	43	8
Index	Index	12	2
			100

The simile is not very good because the preserver of game was more the landlord than the farmer, who frequently suffered depredations from pheasants and partridges he was not allowed to touch. But this is a dangerous field to enter. In his early days, Lloyd George was once campaigning for the Liberals and indicted the Tory landlords as miscreants. He maintained that their carefully preserved pheasants, which under the game laws the tenant farmers were not allowed to shoot, could be seen eating those same farmers' turnips – gross injustice. Alas for the Welsh Wizard! He was refuted by a reference to *Farm Insects*. The pheasants actually were eating the grubs of the gall weevil, often a bad pest, and thus were benefiting the farmers.

The purpose of the book can be seen in another quotation:

To augment his produce has ever been the laudable object of the farmer, and my endeavour will be to supply him with such information as may tend to show him how to diminish his losses, by directing his attention to the general economy of such insects as are likely to injure his crops. (p.x.)

Curtis also makes the point that a careful description of the various pests is important and to read these is not a waste of time for the practical man, as had frequently been put forward.

It is a great mistake to suppose that scientific descriptions and correct nomenclature ought to be employed for the use of those only who are specially engaged in the study of natural history. If insects be not thus accurately and scientifically described, and their names carefully learned, the facts noticed by practical observers are generally worthless, and may tend to mislead, by the confusion of one species with another, and the consequent adoption of improper remedies. (p. xiii.)

Many instances of this are given in the course of the book. For instance, a Mr Middleton, of Middlesex, did not know whether the ladybirds were the parents or the destroyers of the black aphids so injurious to his beans. Another such case was that of the little yellow bodies found on many cabbage white caterpillars, which were not the eggs of the creature, as many people thought, but the cocoons of a parasite of great use in keeping down the pest's numbers. Even scientists could be wrong. Dr Pearson at one time believed the white spots sometimes found on more than half the turnip seed were the eggs of the turnip flea beetle (which Curtis knew well enough were laid on the young leaves). Another of his rivals 'Rusticus' (actually James Rennie) so strongly insisted on this that seeds steeped in brine were sold to insure the grower against the attack of the fly, a remedy which can have been of no value at all. Curtis thought the white dots were fly excrement, pollen grains, or even salt crystals from the brine treatment and he was probably right. It is interesting to note that dressing of cruciferous seed today with powerful systematic insecticides controls the flea beetle as the chemical is transmitted to the cotyledons and renders them insecticidal for long enough, until the main leaves, which the beetles do not like, have grown. The *Gardener's Magazine* (Vol. VI, p. 500) also erred, as Curtis points out:

. . . figs. 93, b and c are there given as the larva and pupa of *Elater segetis* (viz the *true* wireworm) but they are undoubtedly the offspring or produce of some insect not belonging to the same order, but probably to the *Diptera* or two-winged flies. (p. 153.)

Curtis acknowledges the importance of many foreign entomologists or '*savans*' as he incorrectly calls them:* Ray (England), Linnaeus and De Geer (Sweden), Réaumur, Guérin Méneville, Bazin, Bouche and Hespin (France), Harris and Fitch (USA), Passerini (Florence), Köllar (Austria) and his British friends, Marwick, Kirby, Professor Henslow, A. H. Haliday, J. F. Graham of Cranford and finally an anonymous 'Lady in Surrey'.

* 'The misapprehension of the obsolete Fr. spelling *savans* of the plural has given rise in English to the incorrect form *savans*' OED.

This last was possibly Lindley's daughter; Lindley was editor of the *Gardeners' Chronicle*, not the rival journal that made the error about the wireworm.

The pattern adopted by Curtis, as we have already stated, is to deal with pests crop by crop. He starts off with an introduction, noting the importance of the crop and the views of various authorities on it and its pests. He then describes all the insects known to feed on that crop, the known parasites and predators, where found, foreign views and the steps that can be taken against the pest. If the creature has been described in *British Entomology* naturally he gives the reference to it and of course he illustrates the pests, using new plates, in spite of possessing some from his earlier work which he could have used.

The remedies available were of the scantiest at that time: in fact the word 'available' is not quite correct, for one of the most powerful insecticides known – benzene hexachloride (BHC, Lindane, or hexachlorocyclohexane) – had actually first been made thirty-five years previously, by Michael Faraday, in 1825.[32] It took the world's chemists until 1940, that is 115 years, to discover its insecticidal use. We should consequently say the remedies known and proposed were of the scantiest, but Curtis put forward very sound suggestions, using what today we would call integrated control, that is a combination of mechanical, chemical and biological methods.

The first five chapters of his book are on the pests of turnips, as has already been mentioned, and we can get a good idea of the processes available and of Curtis's approach by taking a short look at the remedies proposed in Chapter 1. The worst, or almost worst, turnip pest was the flea beetle or fly (*Phyllotreta* spp), eight are known and six are common. Curtis knew two – calling them *Altica memorum* and *A. concinna*. He suspected the existence of a third species. There was a melancholy song, probably coming down from that age, still being sung in the 1920s, 'The fly, the fly, the fly be on the turmets.' It was a devastating pest and could wipe out sowing after sowing of *cruciferae*. After a careful description of the insects (two species) and their life histories Curtis proceeds to examine the economics, the remedies and the alleged remedies, for he says that 'the recommendations suggested by many eminent agriculturalists . . . have been greatly at variance'. He is not going to 'determine such practical points of difference' and he hopes they (the eminent agriculturalists) will go on working until the variations are explained. He then recounts various procedures and comments on them. He does not think ammonia from manure much affects the beetle. Burning is the best preventative, except on sandy soils; it destroys many chrysalids; presumably he means burning the stubble before sowing, for the Norfolk four course rotation would have been used – roots, barley, seeds, wheat. Feeding off turnips by means of sheep reduces the attack, not so much from the dung and urine as from the consolidation of the land imprisoning the larvae. He does not have much faith in Mr Sutton's 'plan of preparing the fallows for seed, and leaving the land ten days or a fortnight before sowing,' and thinks his success was due to a dry season.

Deep ploughing, as practised by Mr Cowdry, he favours, as it will bury the chrysalids too deeply for emergence. However, he does point out that this is to assume the chrysalids are in the soil; they may be elsewhere as well and thus more information is needed.

Mr Linton's remedy of thick sowing (6 to 7 pounds of seed per acre instead of the usual 3 to 4 pounds) together with a manure giving off much ammonia, he merely reports, and adds that steeping the seed in water and using young seed to get quick germination and growth gives the crop a good start over the beetle.

Dusting with quicklime, hot-slaked lime, soot and ashes gave varying results.

Substitute crops seem to be a possibility in badly attacked areas; such a substitute could be the mangel-wurzel (not cruciferous and not liable to attack) and the Swedish turnip (or swede) which is less attacked. Curtis also thinks that the insects are attracted by scent. If a turnip field had a belt of mangels planted round it, some protection from attack would be given, enabling most of the seedlings to reach the rough leaf stage, when they are safe.

Destruction of cruciferous weeds, such as sauce-alone, hedge-mustard, and common ladies-smock had been advocated by Curtis for twenty years,

> to which these *Alticae* are so strongly attached, for they grow in abundance in every hedge and meadow; they appear long before the turnips come up, and attract and give support to the parents of the future swarms that are to sweep away the crops of the farmer.

Curtis thinks that rolling is of benefit, especially in damp weather or after heavy dews, as many insects fall on their backs when disturbed, stick there and exhaust themselves as they try to recover. Watering, if it can be done, is also good for the same reason, but irrigation is not, for it merely floats the insects off the leaves to which they later return.

Curtis next passes to 'direct remedies'. The 'Paul net' he thinks would be useful even though it is considered to be but a toy. This was a net on a triangular framework with two curved projections out from the base enabling it to be pushed along the ground, where it disturbed and collected insects. It is illustrated in Kirby and Spence's *Entomology, plate XXIV, fig. 3*. Newly painted or tarred boards dragged over the crop collected many beetles during fine weather, but not in the cold or wet when insects are disinclined to move. White paint was best, as the insects were attracted to it.

Burning weeds and stubble to windward has proved effectual, he says, but thinks it would be better to burn the material on the lea side so that the smoke drifting down would mask the smell of the turnips which attracts the insects up wind of the crop. In the light of modern knowledge this is a bit anthropomorphic: it is doubtful if one smell can mask another for an insect, for their sensers pick out and react to a stimulus whatever else may be present. This smell can only be over-ridden by a yet stronger repellant stimulus.

Dusting with road dust at night had proved effective in Hanover. The reason for this is that inert dusts of a granitic nature, such as road dusts tend to be, scratch the protective cuticle of insects by the sharpness of the particles, causing them to lose water by evaporation, and to die. Curtis dismisses many nostrums for treating the seed and dusting the plants, – snuff, assafoetida, ammonium carbonate, etc. – and comes to the need for more information and experiments:

> Such are the remedies proposed; but I fear it is not by the experiments of a few philosophic men that we can hope to discover any positive antidote to so great an evil. We want correct data from every sort of soil under the various influences of climate and effects of cultivation before we can fairly grasp the subject. Until we become acquainted with the *economy* of the beetle, we are groping in the dark. (pp. 32–3.)

Curtis continues with instructions for making pot experiments and then on the importance of preserving birds, particularly the grey and yellow wagtails, as they consume so many pests. The chapter then has the admirable summary and a piece of advice as important today as it was then. A bad attack of a pest is usually followed by a respite of several years and the threat is forgotten:

> thus after a lengthened period of tranquillity . . . they suddenly make their appearance, and take us by surprise and at advantage. The intelligent farmer must therefore be up and stirring, to detect the first breath of infection, and be instantly prepared with his remedy.

It is a little ironic to think that the remedy, BHC, all unknown, was standing in Mr Faraday's laboratory (see above p. 83) a couple of miles down the road, at the Royal Institution in Albemarle Street, and continued unknown for another eighty years, when it burst forth, in 1940, providing an important remedy against this pest, and considerably helping food supplies in a hard-pressed, wartime Britain.

Curtis's chapter ends with the magnificent Plate A showing the two turnip flea beetles, an infected leaf, larva, chrysalids and details of antennae and mouth parts.

It is to be hoped that the above description will lead readers to turn to the book themselves, but in any case we hope it shows how he combined the scientific and practical. He always had his feet on the ground; science for him was to be made to serve mankind and not to be just the amusement of the philosopher. He was a man ahead of his time and put forward views, strange then but accepted today in the post-Carson era.

After the success of the chlorinated insecticides (BHC, DDT etc.) in 1940, and until quite recently, most pest-control technicians faced with a problem would merely look around for the best chemical method of control, but today they are going back to Curtis, albeit unconsciously, and looking at all methods, mechanical and biological included. The latter are much now advocated on the grounds that they cause no pollution, which is true but is not the main justification for using them, as Curtis would immediately have seen. Biological pest control costs almost nothing compared with chemical sprays and it is this rather than potential pollution that appeals to the farmer. Curtis was well aware of the economic side of pest damage, (see p. 88). He collected quite a body of figures, scattered throughout his book. There is, for instance, the case of the turnip sawfly (*Athalia colibri* Christ.; Curtis called it *A. spinarum*). The black caterpillars of this pest had a number of other names, such as black palmer, black canker, black jack, black slug, nigger and negro. The sawfly, almost unknown today, caused more damage then than the flea beetle. A Mr Marshall recorded that the insect destroyed two thirds of the turnips in Norfolk in 1782. 1818 and 1833 were bad years and 1835 was *the* 'Canker-year' and was disastrous. In the south of England seven-tenths of the second sowing (the first having been destroyed) were lost and in many cases a third sowing was also destroyed; and so the accounts continue. Strangely enough, turnips left unhoed did better than those hoed. Curtis discusses the cost of control by hand-picking the caterpillars off the leaves:

Mr Sells states that a boy ten years old gathered the caterpillars in a field suffering in a slight degree, at the rate of 180 in an hour; eight hours *per diem* would give nearly 1,500, or 9,000 a week; so that ten or twelve children, from six to ten years old, would collect 90,000 or 100,000 in a week, where they are *not* abundant; in such a case 6d. a pint, and 2s. a day to the superintendent, would probably answer the purpose. (p. 57.)

It will be noted that this is quite a realistic figure, three a minute, and the pay quite good for that age, for the children could probably earn at least 1s. 6d. per day or 9s. per week. They would be the farmer's and farm labourer's children and thus the cost of pest control was kept within the agricultural community and provided the poorly paid workers with additional income, rather than going outside, as it would today, to the pesticide and machinery manufacturer. One does, however, in passing, feel a little sympathy for the 2s. a day superintendent, attempting to control his gang of a dozen six to ten year olds for 12s. per week.

Curtis puts great faith in a form of biological control – the use of young ducks which have been trained to feed on the larvae. Why the insect should be so rare today is a difficult question to answer. Probably most of them were migrants and as considerably fewer acres of turnips are now grown the critical minimum population has been passed. The immigrants fail to find suitable plants on arrival in Britain.

Though not strictly speaking economic, we feel one should comment here on Curtis's efforts to propagate scientific truth. The sawfly larvae usually appear with great suddenness and the idea grew in many parts that they fell from the sky. Curtis admits that aquatic animals may fall from clouds with rain, having been caught up in whirlwinds and tornadoes and he quotes cases of the fall of fish and frogs, but finds 'that the negro caterpillar should *always* fall in turnip fields would be very extraordinary'. (p. 38.) These insects came from eggs and the eggs from sawflies, so the idea was nonsense. A little earlier he said:

It is to be hoped that the veil of superstition has long since departed with our ancestors; but it is still necessary to dissipate the clouds of error which obscure the beauty of truth: this is the pleasing province of the naturalist, especially of the entomologist. (p. 37.)

He also issues warnings that substances distasteful to man, such as bitter herbs or assafoetida are not necessarily so to insects: he was against anthropomorphic reasoning.

Let us now return to economic considerations. Curtis records a number of direct observations as, for instance, the attack of a small black weevil (*Apion apricans*) on the seed clover crop. He gives the following figures:

				£	s.	d.
1798	4.5 acres produced	16.75	bushels worth	41	17	6
1800	,, ,, ,,	7.5	,, ,,	18	15	0
	Loss	9.25	,, ,,	£23	2	6

He was also well aware of what the economist calls 'the sacrificed opportunity'; for instance, he reports that many farmers were obliged to stop growing oats because of wireworm attack and they put the fields down to pasture, a less valuable crop. He was against the uneconomic use of land, as is instanced by his remarks on the potato blight. The disease devastated potato crops from 1844 onwards and caused much suffering, as has been so well described by Mrs Cecil Woodham-Smith.[90] Curtis is not sure of the cause of the blight, but with simple logic is against the view that it was due to an insect, as so many maintained, such as A. Smee in 1846. Curtis seemed to be moving towards the view, put forward by M. Guérin-Méneville in Paris, that the disease was caused 'by a cryptogam', as was indeed the case and positively stated to be so by the Reverend M. J. Berkeley.[3] Curtis was against wasting land, and thought potatoes should not be much grown until a cure had been found. He says (writing in 1849) but reprinted in *Farm Insects*):

The very great breadth of potatoes planted in 1848 has given so much larger an amount of produce, that the loss from disease is not felt as it would have been, had the usual quantity (i.e. area G.O.) only been grown; and this is a very important fact to keep in view, as regards our future prospects. Indeed is it worth while *at present* [Curtis's italics] at least, to sacrifice so many acres of valuable land in growing rotten potatoes?

Here is seen another view of the nature of losses due to pests; it is the area of land sacrificed to them, in effect set aside to feed fungi, viruses and noxious animals, chiefly insects. To get something like the quantity of potatoes needed by the market, farmers were growing a considerably greater area than before the blight struck, drawn to it by the increase in potato prices. The difference between these two figures is the area used to feed the fungus *Phytophthora infestans*. It would have been much better to grow substitute crops – a mechanical method of pest control. (Appendix 6.)

Curtis was also something of an ecologist before the word was invented,* as he was well aware of the interaction of the different life-forms in an area. For instance, in discussing the bad attacks of the pea and bean beetle (*Bruchus pisi* for Curtis, but now known as *B. pisorum* L.) he points out that the greater scale of attack in the chalk areas of Kent may be due to the presence of wild flowers attractive to the beetle (p. 366). He also notes that much pea and bean seed is imported, nearly all of which is considerably infected with this pest, a case for quarantine. Another similar ecological observation was in the matter of the turnip sawfly (see page 86); the larvae ate only the young leaves and appeared to leave the older leaves to nourish the plant, thus not utterly destroying the plant on which they depended for food.

As regards quarantine we mention here his warning on a case in Barbados, noting though that it was in his *Gardeners' Chronicle* articles and not in his book. After a hurricane, coconut trees were brought in as replacements and introduced the coconut scale which did considerable damage. Curtis suggested the use of sulphur sprays and the introduction of a predacious beetle, only known in Europe – in other words 'integrated control'.

In all, *Farm Insects* was a remarkable book and set the farmer and administrator on a course of raising agricultural production

* The translation of Haeckel's *History of Creation* (1873) is the first use of this word in English (but spelt 'oecology').

by means of pest control, even though the work's circulation was so limited – 2,000 copies. At £1. 2s. each we get a total of £2,200, of which Curtis would have got 10 per cent or £220, not highly profitable but a welcome addition to his income. It must be kept in mind that he had already been paid for the articles forming the basis of the book by the Royal Agricultural Society (see page 73).

The book, and the RAS articles, were in such demand that long after his death in 1861 a new edition was put out by Van Voorst, in 1883.[17] Little had changed in the pest situation between the editions and the latter book is merely a reprint of the former with a short introductory 'Advertisement' pointing to the success of its objectives. It at once

gave an impetus to the study of economic entomology by the British agriculturalist. The importance of this subject has been so fully appreciated in the United States and on the Continent, that there now seems to be a general disposition on the part of our farmers to learn more of our Farm Insects.

As to the objectives, they are best described in Curtis's own closing words:

The first step towards vanquishing an enemy is to ascertain correctly his habits; the next, to be so certain of his appearance as not to mistake one party for another; and a third and no less important object is to be well acquainted with our allies and friends. For want of this it has often happened within my own knowledge . . . that the very insects ordained by the Creator to keep noxious insects in check have been mistaken for the offenders; and thus the most serviceable auxiliaries have been persecuted, and even sacrificed to our ignorance of their deserts. (p. 514.)

This was still true during the insecticide bonanza of the 1950s, not so much that we did not distinguish between friend and foe, but that we did not think it important enough to worry about; in the same way that, sooner or later, in all wars generals start shelling and bombing their own troops. It is just one of the risks of the operation. Consequently we persecuted our auxiliaries by drenching everything in sight with pesticide. It resulted in the creation of even greater attacks and the formation of resistant races of pests. We now know better.

The above quoted portion continues:

I cannot but hope that I have been permitted to lay a foundation of the insects injurious to man on which a more sound and perfect superstructure may gradually be raised, as materials are collected to increase and correct the mass of information which I have gathered together. (p. 515.)

Curtis continues by saying it is most important that the agronomist, 'however talented he may be', calls in the aid of the scientific entomologist in order to get correct information. He is advocating what today we call the 'interdisciplinary approach'. His charming *envoi* to the book is typical of the man.

I will now bid farewell to my agricultural friends in the good old English phrase, 'May God speed the plough!' I sincerely trust that my labours may tend not only to the instruction and to the benefit, but even to the amusement, of those engaged in the cultivation of the soil; and that the first step towards a correct knowledge of economic entomology may induce others to take up this important subject. After forty years service in this labour of love, it is time for me to relinquish my pen and pencil, and release myself from my toil, which I cannot do better than in the words of the Mantuan bard, who delighted to sing of rural pursuits:

> For, overlaboured with so long a course
> 'Tis time to set at ease the smoking horse.[81]

It would have been a great satisfaction to him to know that his plea was heeded and many others were induced to take up this important subject; some of the earlier ones we shall notice in Chapter 10.

7A

9 Pest Control and Maturity

In the previous chapter we discussed *Farm Insects*, which, though published in 1860, was compiled between 1841 and 1857. We now need to turn back to 1851 and consider Curtis's mature years. There are no letters in the Haliday correspondence at the Royal Entomological Society between 11 May 1846 and 13 March 1862, but this does not mean that the two friends were not writing to each other, as Curtis expressed great delight in the former letter at having heard from Haliday after a considerable time. It is more than likely that some letters are missing. Consequently, apart from his professional activities, not much is known about Curtis from 1846 to 1860.

1851 was the year of the Great Exhibition of the Works of all Nations in Hyde Park, the triumph of the Manchester school that heralded a new age. It does not seem to have attracted the entomologists, though cochineal and shellac were commodities much used at that time.

In 1851 Curtis became a member of the Entomological Society and was elected its president in 1855, from which important post he resigned in 1857 as he could not afford it. William Spence proposed and H. T. Stainton seconded the motion. Curtis's *British Entomology* appears to have been popular there, for the committee always seem to be arguing about it. For instance, in February 1848 Mr Ingham moved, and Mr Newport seconded, that the work should never be removed from the Society's rooms, a decision reversed a few years later (March 1852), when the borrowing of one volume at a time was allowed. That same month a 'Cabinet Committee' was formed, not, one hastily adds, an inner Cabal designed to run the society, but a body to organize its collection of insects. Insect cabinets were vastly important in those days. An interesting point about this committee is that it included both Curtis, Westwood and J. F. Stephens, at one time Curtis's great rivals and *bêtes noires*. Perhaps they were reconciled by this time, or the presence of two other members, Messrs Wing and S. Stevens, kept the peace.

His 'professional' publications, as distinct from the pest control ones, covered a wide field; for instance, sixteen new species of Tortrix moths 1850, an apparatus to destroy mould on insects in collections, a new sawfly on lily-of-the-valley, and a large number more, the titles of which will be found set out in his French obituary notice.[87] He can be seen in all the important journals, the Entomological, Linnean and Zoological Societies publications, the French entomological society, the *Zoologist*, the *Entomological Magazine*, the *Annals of Natural History*, the *Naturalist* and many more.

We have not yet discovered when Curtis's first wife died; possibly it was before Christmas 1855 for he spent the festival that year with Miss Lindley and her brother; in his letter of thanks he reminds the lady of their early days together in Norwich and encloses an autograph for her friend's collection. At this time he was living at 18 Belitha Villas, Barnsbury Park, with his son Edward.

We have already mentioned the absence of Haliday letters between 1846 and 1862, but a little about Curtis the man and where he lived can be gleaned from his entomological diary. This diary is a large octavo book of blank pages, and of good paper, bound in leather and now with the Royal Entomological Society, London. Its leaves are faintly ruled with horizontal lines and have three columns made by two red vertical lines in the left hand margin, as if for £ s. d., but on the left instead of the right: probably they were especially printed for him. There is a page for each day of the year, except February 29th, and each day contains a number of observations on plants, insects and other animals, written in Curtis's most minute, but reasonably legible, hand. The first column is for the year, the second for the host (apple, man, wheat etc.) and then the observation follows. The diary is thus like a 'five-year diary', the events on the same day of successive years being recorded on one page. The observations are very brief:

January 28, (heading) 1846 House. *Culex annulatus* on window of sleeping room.

But the curious thing is that the years are not successive. The first entry may be, say 1848, the next 1836, the next 1849 and so on, a complete jumble. One presumes the records for the earlier dates on lower lines were put in from note books and correspondence as and when Curtis had time, in order to get a consensus of observations on insect behaviour on a recurring daily basis.

The observations on natural history, mostly insects, are fascinating, but it also is interesting to go through this diary to see if, firstly, we can add anything to our picture of Curtis the man, and secondly to find out when he left Hayes to return to London. A difficulty in both cases is that Curtis seems to have been an immense traveller. Obviously his opinions were widely sought and he may well have received fees both for his help with insect collections and for his advice on pest control. The observations which can definitely be placed at Hayes or London are not very many. He appears, for instance, to have been giving advice to the queen, or her gardener, at Frogmore; her lavender had been attacked by a cuckoo spit insect. He is often off to Glanville Wootton and further afield too, for in 1853 he records that he saw the first swallows on April 8, heard the first cuckoo on April 10 and the nightingale on April 17, all at Pau. Specimens and letters were sent to him from all over the world.

As far as one can see, he left Hayes in 1849, probably at the Michaelmas quarter, for there are records of various insects in the house and privy and on the hops Curtis was growing in his garden. In 1846, for instance, he got three-quarters of a pound of dried hops in spite of aphis attack. On 14 April 1850 he records '*Anacacup sarcitella*' male in a vinegar saucer in the 'sleeping room' at Belitha Villas, London. We note a curious distinction between a 'sleeping room' and a 'bedroom', the latter being mentioned more often in London (in spite of the insect in the vinegar saucer mentioned above) than in the country. In London he has a water closet, at Hayes a privy in the garden.

A few items from the diary throw a little light on what he and his wife (if she were still alive) were like. He (or they) might be said to like flowers in the house, for he found aphids there on winter roses on 16 February 1845. He was always so anxious about his hops that one might suppose the Curtis family made some of their own beer, and possibly they drank gin and wine for he records 'swarms of *Drosophila cellaris* floating in an unstoppered gin decanter' (16 May 1849) and *Tinea* larvae in corks of wine bottles (19 May 1845). They kept a dog, for on 12 August 1850 he found thousands of *Pulex canis* eggs where the dog slept. Honeysuckle grew round the door of his cottage, yet though but a cottage he had a study and whether his 'sleeping room' and 'bedroom' were one and the same or were different rooms it is difficult to say. From the above and from the name – Hayes Court – it hardly sounds like a cottage. In any case he did not like it: it was damp, the walls were thin and did not keep out the cold. He thought he had been cheated over it and obviously he was glad to get back to London again.

Farm Insects, as we have already noted, was but a reprint, with a few emendations, of his papers (1841–56) prepared for the Royal Agricultural Society and these last must have considerably helped his income. Fortunately the minutes of the Society's Finance Committee still exist, both in the 'Fair' and 'Rough' state. From the payments made to John Curtis we can conclude that the society thought very highly of him, for they are considerable. For instance: £72 odd (4 December 1844) for an article on turnip pests (the fifth of the series) is about as much as one could expect to get for a similar account today. The main payments to Curtis so far traced are given in the table shown on page 96.

A quick glance at the above sums suggests that they are very exact figures and one is thus tempted to work out a wordage rate (so much per thousand words) that Curtis was getting. A second look, however, shows that some of the items are guineas. For instance, the first entry £25 4s. od. is twenty-four guineas; that for 4 December 1844 is sixty-nine guineas. Gentlemen were paid (if paid at all) in guineas and journeymen in pounds and we cannot say how the totals were reached. Most likely by discussion under the Old Chums Act, as now.

These figures do suggest that Curtis was now making the extra £50 or £100 a year that he felt he so needed.

In 1858 his sight began to fail; he had always strained it very much in dissection and engraving work.

On the 27 June 1860 John Curtis married, at St Mary the Virgin's Church, Matilda Bliss, aged thirty-two, the daughter of Samuel Bliss, a law stationer, living at 12 Stanmore Street, Caledonian Road. Curtis was described as 'widower, annuitant and gentleman, the son of Morgan Curtis, gent'. The whole family had risen with the passage of time, a not uncommon event in all ages. Stanmore Street is in the Barnsbury Park area, so one assumes the couple met there, possibly at this same church. Matilda must have been a woman of some force and character

Entries in 'Fair' book of Finance Committee Minutes of the Royal Agricultural Society, London

Date	*Entry*	£	s.	d.
2 June, 1841	Mr Curtis for Report on Insects injurious to the Turnip crop, with original drawings	25	4	0
	Mr Curtis for engraving and printing plate for *Journal* on the same article	23	12	0
5 January, 1842	Mr Curtis's account for Literary work and engraving was recommended to be paid on his delivery to the Society of the original copper plates employed			
	Mr Curtis	32	11	0
	Engraving and printing plates for *Journal*	23	12	0
3 August, 1842	Mr Curtis, illegible (probably 'assessed') Bill	101	7	0
	On account	6	0	0
7 December, 1842	Mr Curtis, engraving £89 6 0			
	Deduct 6 0 0	83	6	6
3 May, 1843	John Curtis, engraving	112	7	0
7 February, 1844	Mr Curtis for report and engraving	74	3	6
4 December, 1844	Mr Curtis, report	72	9	0
6 August, 1845	*Journal.* Mr. Curtis's Report	61	11	6
2 December, 1846	Mr Curtis, paper for insertion in *Journal*	76	5	6
3 February, 1847	Mr Curtis, on account	60	0	0
14 April, 1847	Mr Curtis, balance of account	58	19	0

to take on a poor, blinding, elderly man, thirty-seven years her senior. It was by no means a companionship or a paternal relationship marriage, at least we hope not, for their first child, a boy called Robin, was born on the 14 April 1861, and a second son followed about a year later, named Henry Alexander after his godfather A. H. Haliday.

John by now was completely blind. Two days after the birth of Robin, his pension was increased to £150 per annum on account of his affliction. The 16 April 1862 must have been a happy day for him.

From the marriage onwards, and possibly before, Matilda wrote Curtis's letters (this may have been how they met – Curtis seeking an amanuensis from the nearby law stationer, Samuel Bliss) and they are much more difficult to read, some of them even being crossed. The newly married couple went to live with Edward Curtis and his wife Emma at Belitha Villas, but Edward and his family had moved away by the time of the 1861 census; this shows John Curtis 'gentleman and annuitant, aged sixty-five'. (He actually was seventy and probably had taken off five years during his courtship of Matilda). As, owing to his failing sight, she would fill in the census form (the census was usually held at the end of March) he could hardly blurt out the horrid truth just before her first child was born. Matilda Curtis was entered as aged thirty-three and there was a Susan Goodlett, twenty-two, doubtless a servant. Though Curtis's health was failing he was very happy in this different life with his new family, both on the technical and domestic sides, as the letters to Haliday show. On 13 March 1862 Curtis had received from Paris a book on pests by Coln. Joureau, called *Les Insectes nuisibles aux Arbes frutiers, aux Plantes potagères, aux céréales et aux Plantes fouragères*. It acknowledges the debt due to Curtis and *Farm Insects:*

> but I fear from what I have had read to me that it is little better than a compilation, now what is wanted are original facts regarding the economy of different insects in the different countries of Europe in order to ascertain the extent of their ravages, the periods of their appearance, and all the parasites connected with them.

This letter suggests that Matilda was educated and at any rate was able to read French, but was not very familiar with the language for, in writing it, she leaves off the accents.

Curtis continues this letter saying he hopes Ghiliani is better and he sympathizes with him for he himself suffers from 'neuralia' (*sic*, 'neuralgia') or rheumatism in the abdomen, or gastritis and is harassed day and night. He had received an enjoyable visit from Sir Charles Lyell who was preparing a work on the *Antiquity of the World* and J.C. says it contains no evidence of the existence of man earlier than the pyramids of Egypt and the ruins of Stonehenge, Nineveh and Carnac. One might think Curtis was moving towards an acceptance of Darwinism (*The Origin of Species by means of Natural Selection*[23] had been published in 1859 and owed much to Lyell. The latter's book *Principles of Geology* (1830) had reached young Darwin at Montevideo when he was on the *Beagle*. It influenced his whole thought for ever afterwards.) One does not know how Curtis reacted to Charles Darwin: they were both members of the Entomological Society. Westwood bitterly opposed Darwinism: perhaps Curtis reserved his opinion. Curtis ends this letter saying his wife is in good health and sends her remembrances and that: 'little Robin is very jolly, eleven months tomorrow and making good use of six teeth'.

A little later a letter dated 2 June (presumably 1862) reports further ill health and a trip to Margate to try and effect a cure. It was of no use; he was living on spring water and had lost a pound in weight a day for two to three weeks. He warns Haliday about buying copies of *British Entomology* from Lovell Reave without examining them first, as out of 500 supplied half were imperfect. This must have been very distressing to Curtis, the perfectionist. One presumes the examinations must have been made by Matilda. Curtis ends his letter on a more cheerful note: 'The little boy would send you a kiss if he knew how to do it.'

In writing to Haliday on 26 August 1862 Curtis thinks his end is near, or at least will be 'in the course of a few years' as he felt a little better and asks a favour of his friend, that he will be the new baby's godfather, to which Haliday agreed. The name Cornelius was also added.

John Curtis died on 6 October 1862. On 17 November 1862 Matilda wrote to Haliday saying he must have seen the news in the paper. In this letter she showed her great love for her so different husband. She first tells how he was recovering, but his old complaint returned, causing him much pain day and night and this lasted twenty-one days and exhausted him. On 6 October,

> he resigned his valuable and sad life of blindness and suffering to sights of glory and everlasting joy for blessed are they who rest in the Lord . . . Such were the constant expressions of my dear husband. The sweet heavenly smile that lighted his countenance when he departed [showed] that he is released from all care and suffering . . . I deeply feel the loss of this kind, affectionate and intelligent companion.

Curtis also must have had great faith in and love for his young wife. In an age when a gentleman regarded women as inferiors, dear, decorative, delicate, little things (unless they were charwomen, agricultural labourers or miners) who had to be guarded and protected, Curtis, in his will makes Matilda his sole heir and executrix, finding no need to hedge her around with trustees and expensive legal luminaries.

Curtis's body was submitted to a *post mortem* and death was found to be due to stomach cancer.

Matilda had not only lost a kind, affectionate and intelligent companion, she had also lost an income. With two young children to bring up, she faced her difficulties with considerable skill and fortitude. She moved out of Belitha Villas to a cheaper house, No. 7 Stack Orchard Street, and soon afterwards had to move again, because the house was wanted for the railway. Eventually she settled at 18 Huntingdon Street, Barnsbury Park, the continuation of Belitha Villas westward on the other side of the Hemingford Road and just above some pleasant trees and grass – Thornkill Crescent and Gardens – where one imagines her children later would have played. She approached Sir Robert Peel for a continuation of some of her husband's pension and sold his books, and his insect collections, to the highest bidders. The highest bidder for the insects was Professor MacCoy, of the Natural History Museum, Melbourne, Australia, where they are much prized and may be seen to this day. Matilda received £500 for the main collection and £100 for a subsidiary one.

The books were sold by auction by J. C. Stephens at No. 38 King Street, Covent Garden, on 8 June 1863 'at half past twelve o'clock precisely'.[78] The time throws an interesting light on social habits. Most sales today stop at this hour to enable the customers to go and eat. The sale catalogue still exists (in the British Museum) and throws an interesting light on Curtis's character. There were some 700 volumes in the sale: presumably his widow would have retained a certain number for herself. The vast majority of the books are on insects and the list covers all the standard works from Mouffet's *Insectorum Theatrum* (1634) onwards. There is scarcely one of the ancient entomological writers missing and the list turns the modern insect bibliophile green with envy. Dejean, Denny, Donovan, Cuvier, Fabricius, Geodart, Harris, Latrielle, Linnaeus, Meriam, Réaumur, Schkur – you name it; Curtis will have had it. He also had the available works on pest control: Major's book on fruit tree insects (1829), Harris on the *Insects of New England* (1852) and

Fitch's *Noxious Insects of New York* (1856). There is also quite a collection of botanical and other natural history books, including the now much sought for Audubon *Birds of America*, and works on chemistry, the microscope and agriculture. These were the tools of his trade. The small residue of other books tell us something of him. In religion, for instance, he was no bigot, for we find a copy of the Koran, Hawker's *Poor Man's Concordance* and the 'dangerous' *Evidences of Christianity* (1804) by Paley, which he appears to have purchased in spite of Haliday's warning. Or perhaps, then, as now, warnings only make such works more popular. Owing so much to Kirby, he naturally had his *Power and Wisdom of God* (1835) as well as his entomological works.

Also sold was Durham's *Physio-Theology* (1798) in two volumes, with plates, whatever that work might have been. He was interested in churches, too, and had books on Ely and Salisbury cathedrals and runs of magazines on Norfolk rivers and Cotman's *Engravings of the most celebrated Brasses in the county of Norfolk*.

Being so interested in Italy he naturally had Forsyth's *Italy* in two volumes and a few other travel books as well. There were some books on painting and design, such as De Pile, *Art of Painting*, Da Vinci, and Hoare's *Arts of Design in England*, among others. He played chess, it will be remembered, so he had Philidor's two volumes *Chess* (1810). Concern about his health may be seen in a few books on this theme – Buchan's *Domestic Medicine* (1826), perhaps merely the equivalent of the first aid book of today, but a deeper interest in the subject is seen with Lot 23 *The Diary of an Invalid* (2 vols), *Instructions on Health* and *The Lady of the Lake*, showing that strange mixture of subjects beloved of book auction catalogues to this day. One wonders whether Scott perhaps gave him the copy of *The Lady of the Lake* when he visited the poet so splendidly accoutred in Edinburgh.

Among the general works we find Burton's *Anatomy of Melancholy*. Had it, in view of Curtis's religious readings, led him to think that 'one religion is as true as another'? He had the great Gibbon's *Decline and Fall of the Roman Empire* expressing somewhat the same view too. The catalogue ends with an unusual penultimate item, No. 261, strange for Curtis; it is Hieronymus Brunchwygk's *Buch gennant 'Liber de Arte Distillandi*' (1500). We thought that Curtis may have made his own beer; if he had been thinking of making his own gin he would seem to have been using a somewhat out of date guide! The final item, No. 262, is 'a folio bundle of various tracts and two others'. We do not know how much the sale made. Contemporary secondhand dealers' catalogues for this type of book quote prices from 10s. to £1 each, so Mrs Curtis may well have received some £350 for the 700 volumes and perhaps more.

Westwood also purchased Curtis's drawings used to make the plates for *British Entomology*, some manuscripts and the diary already mentioned. The drawings are now in the Natural History Museum, London.

Curtis was universally mourned and many obituary notices appeared, the longest and most complete being that in the *Annales de la Société entomologique de France.*[87] It was by I. O. Westwood and is mostly a translation of the obituary by the same man in the Linnean Society's *Journal*[86] but with additions. Westwood, whom Curtis regarded as a rival at one point, says of him in both French and English:

> [He] was very reserved in his communications with such of his brother entomologists as were known to be engaged in works intended for publication; but to those with whom he had unreserved intercourse his manners were engaging and kind.

Westwood also quotes a letter from Haliday in both obituaries, at greater length in the French publication. Haliday says he was most upset by Curtis's death; he had known him for more than thirty years without friction of any kind and says:

> he was indeed very loveable, warm-hearted (too much so perhaps for his own tranquillity), pure-minded and honourable.

This seems to be a very accurate summing up of his character and one any man could wish for.

10 Curtis's Achievement and Succession

Curtis achieved at least two important things. He established that a man could live from the practice of science and he created economic entomology, thus opening the way to modern pest control. His reputation in this field persisted for a long time and indeed up to this day (see page 106). In the 1880s C. V. Riley, who in America developed the same ideas as did Curtis in Britain, still admired him and, apparently, was not aware that he was dead. Riley wrote to Curtis:

> I have contracted to prepare a hand-book of Entomology for Schools and Colleges, based, of course, largely on American species. Yet I shall have frequent occasion to refer to your own works and to quote from your writings and probably use some of your drawings, with due credit. I should esteem it an honour to be permitted to dedicate the work to you and would therefore ask of your permission to do so.
>
> Sincerely yours
>
> C. V. Riley

The letter is undated, but is on Government notepaper, Washington DC, with a line '188–', but the actual date not filled in.

C. V. Riley, an immigrant from England, became the first U.S. federal entomologist. Among his many triumphs were the introduction of spraying against the gypsy moth and Colerado potato beetle in America, and the great help he gave to France in controlling the devastating Phylloxera of the vine.

Curtis did so much to promote the view that science could help pest control that public opinion, or at any rate the opinion of some of those in power, started to change. We have already noted that Curtis received a pension to enable him to carry on his work, what today we would call a research grant. Such government action was bitterly opposed in certain circles, particularly those on the arts sides of the older universities. For instance, Lewis Carroll wrote a poem called *Fame's Penny Trumpet* which was a terrific indictment of scientists who expected to be given money for research. The poem was 'Affectionately dedicated to all "original researchers" who pant for "endowment".'

> Blow, blow your trumpets till they crack,
> Ye little men of little souls!
> And bid them huddle at your back –
> Gold-sucking leeches, shoals on shoals!

He compares them with Plato and Newton, who worked for nothing, and solely to benefit mankind. Plato and Newton were gentlemen and did not need to suck gold, at least not in public, as they obtained it painlessly – the community supported them (see Appendix 7) by the operation of the system.

The Reverend Charles Dodgson seems to have forgotten that he himself was a gold-sucking leech – a mathematics don at Christ Church, Oxford. Probably subconsciously his opposition was economic. If the university income, unlikely to be increased, had to be spread over a lot of new chairs, there would be less all round for the establishment. On the other hand it must be remembered that Cambridge had had a chair of botany since the early eighteenth century and Edinburgh a chair of agriculture and rural economy from 1790.

We mentioned that opinion started to change, but the change was not rapid, or if it were it was not acted upon for some time and when it did come was furthered by a rich amateur – Miss Eleanor Ormerod (1828–1901).

Eleanor Ormerod was the last child of ten of a Lancashire property owner, George Ormerod. His property contained coal – hence their wealth. She was highly intelligent: as a child she had a vague interest in natural history. She took up the study of entomology in 1853 when she was twenty-five years old and, in her autobiography, says that a strange locust found in the village bordering her father's Gloucestershire property ('The extent was not very great, only about 800 acres'!) probably pushed her into it. It became her life's work.

She was a woman of some character. She started to teach herself entomology from J. F. Stephen's book on beetles,[76] a volume with no illustrations, but by using this book, constant study and the dissection of specimens she mastered the subject. She was particularly interested in pest control and greatly admired John Curtis's *Farm Insects*, both for its literary content and the fine illustrations. This work can be said to have inspired her and turned her towards useful pest control activities. Her biographer, as late as 1904, still supported her view of Curtis. In a footnote Robert Wallace writes: 'The author of *Farm Insects* (to this day the most beautifully illustrated standard work in English on the subject) . . .'[85]

In due course Miss Ormerod purchased from the publishers electro plates of many of the woodcuts in *Farm Insects* to illustrate her own works.

In the summer of 1868 there was a great exhibition in Paris. The British had been asked at rather short notice to supply an agricultural exhibit and were encouraged by the *Société d'Insectologie Agricole* of Paris to include a display of noxious insects. Mr Andrew Murray, of the Royal Horticultural Society, undertook to arrange specimens sent to him, and Miss Omerod supplied a large number from the 'not very great' Gloucester estate. Mr Murray and the Royal Horticultural Society were obviously delighted with this lady's help, presented her with the Floral Medal and continued to call on her services.

In 1877 Miss Ormerod became Honorary Consulting Entomologist to the Royal Horticultural Society. In the same year the Royal Society of Arts endeavoured to get the government to appoint a government entomologist, as had been done in the USA. The proposal was made by the Duke of Buccleuch and Andrew Murray,[52] Eleanor Ormerod's *confrère*, but nothing was done. Mr Murray even brought the matter up before the president of the Privy Council, asking them to take action against injurious insects. All they did was to inform Mr Murray that they had sent copies of his letter to the country's Agricultural Societies. Murray concluded: 'The reply is somewhat Delphian in its utterance . . .'. Eleanor stepped into the breach and started publishing, at her own expense, annual reports on injurious insects, which ran until 1900 and were immensely popular and saved the government a lot of money.

Curtis's *Farm Insects* was reprinted in 1880, as we have already noticed; a factor leading to this happy event was the lady's continual praise of the work: it further increased the interest in pest control. In 1882 Miss Ormerod became Honorary Consulting Entomologist to the Royal Agricultural Society which led her to be busier than ever in this field. In 1885 Mr Charles Whitehead put it to the Council of Agriculture that it would be valuable to publish reports on injurious insects and in 1889 he was retained as Technical Adviser to the newly formed Board of Agriculture. With Miss Ormerod's help he prepared some leaflets on pests. Between 1892 and 1900 Miss Ormerod wrote a number of books on pest control, which were illustrated by her sister Georgiana. In 1894 the Royal Agricultural Society realized that the work was getting too much for Miss Ormerod alone (she was sixty-six) and they appointed Mr Cecil Warburton as consulting zoologist. Warburton later became connected with Cambridge University and for some years was employed by the Board of Agriculture, at a salary of £200 a year, to advise on pest control. Economic entomology was also taken up by Birmingham University (Walter E. Collinge) and the South-eastern Agricultural College, at Wye, Kent (Professor F. V. Theobald). Thus at the turn of the century the subject was well launched.

All this sprang from John Curtis's quiet dedication to practical entomology, to his determination to overcome obstacles and to his fine pen, pencil and burin, combined with the happy coincidence of an equally dedicated coal-rich spinster who appreciated his powers and followed his lead. It is perhaps a slight compensation to the men, women and children* who worked so hard on the dark coal seams that their inadvertent support for the Ormerods enabled at least one of them to benefit the country so greatly, and secure the triumph of the work Curtis started. (Appendix 7.)

Another factor leading to the development of economic entomology here was the interest being taken in the subject in the United States, and the use of the early arsenical insecticides there. Miss Ormerod was in constant touch with American entomologists such as C. V. Riley, L. O. Howard and many others. On the occasion of her receiving the honorary degree of LL.D. at Edinburgh in 1900 she sat next to the American ambassador. In a letter to L. O. Howard (30 April 1900) she wrote:

> As I took my seat by him after receiving the degree he gently whispered 'I congratulate you; you did it splendidly,' and I thought it very interesting that my first congratulation should be so kindly given me by the ambassador of the greatly advanced country to which I am so indebted for help in my work.[85]

* It was not till the passing of the Ashley's Act of 1842 that women and children were forbidden to work underground.

Today pest control is turning back to the variety of methods Curtis advocated, and the economies of the subject are receiving increasing attention, just as he would have wished. Moreover, he is not forgotten. An annual 'John Curtis Woodstock' lecture on the economic aspects of pest control is given by the Royal Society of Arts, and is sponsored by the Shell Chemical Company. These Shell lectures succeeded a series of 'John Curtis' lectures (1964–9),[58] started by a private individual and the Royal Society of Arts, which lectures in their turn had succeeded a series of 'Fernhurst' lectures, organized by a commercial company and, of course, the RSA. This company ended the 'Fernhurst' lectures in 1961, presumably on the grounds of economy. One might presume that it foresaw a more 'Curtisian' approach to pest control arising again, with a lesser reliance on pesticides and consequently lower potential profits in that line. The new attitude was not called 'Curtisian'; it was called 'integrated control' and few realize, even today, that John Curtis was the first advocate of this system and should be honoured for it. He was a remarkable man; he wrote so well and accurately. He can hardly be faulted to this day.

> 'Tis hard to say, if greater want of skill
> Appear in writing or in judging ill;
> But, of the two less dangerous is th'offence
> To Tire our patience, than mislead our sense.
>
> *Pope*[62]

Curtis never bores or misleads us; we need people like him today.

Appendixes

Appendix 1: John Curtis Sources

There are four main sources for the early life of John Curtis: (i) The *Dictionary of National Biography* (DNB),[28] (ii) The *Archdeacon's Transcripts* (AT)[1] in the Norwich Records Office, (iii) The *Index to the Norwich Cathedral Monumental Inscriptions, Eastern Daily Press*, various dates, 1917[37], and (iv) *Land Tax Records*, Norwich Records Office[43], and I am much indebted to the archivist, Miss J. M. Kennedy, for help in examining these documents. Minor sources are passing remarks in letters from Curtis or about him.[35]

(i) There is an error in the DNB where Charles Morgan Curtis is stated to be John's elder brother. He was, in fact, four years younger than John. One can see how the mistake arose as there was an elder brother, also called Charles Morgan, who was buried 21 September 1795, aged five. There is no direct evidence for John's birth, but one feels that as the date 3 September 1791 is so often quoted, it must have been known and accepted in John Curtis's circle. The Linnean Society's obituary has the date as 3 December 1791, an error.

(ii) In the *Archdeacon's Transcripts* (the yearly return of births, marriages and deaths) the father is always referred to as 'Morgan Curtis' rather than Charles Morgan Curtis, but the facts fit the story so well that we think Morgan and Charles Morgan must be one and the same man. A curious feature arising out of the transcripts is that the first son called Charles Morgan was not baptized until he was four years old. No risks were taken with the second Charles Morgan, who was baptized at one day old, on 15 October 1795.

(iii) A reporter or contributor to a series in the *Eastern Daily Press* of 1917 indexed the monumental inscriptions in the Norwich churches, Dissenting Chapels, Friends' Meeting Houses and Jewish Burial grounds, and this index supplies:

(a) The first daughter, Frances, born, 1789, died 1792.
(b) The third son, unnamed, who died 'aged fifteen months', no date.
(c) The exact date of Charles Morgan Curtis's (the father's) death, namely 4 January 1796.
(d) The date of Mrs Frances Curtis's death – '30 November 1822 aged sixty-three' and thus the date of her birth and her age when her first child was born.

(iv) As in the case of the transcripts the owner of the house in the Close is in the Land Tax records given as Morgan Curtis and not Charles Morgan Curtis, but here again we think Morgan and Charles Morgan are one and the same person. Robert Owen appears to have rented this house at £3 a year, to have gone away for a year and then to have returned, when the owner trebled the rent to £9, it being reduced again in 1794 when the Reverend Dr Forster took it over.

The clue giving access to all the above is an entry in the catalogue of the Norwich Society of Artists' spring exhibition of 1817. John Curtis showed three pictures and gave his address as 'Lower Close, Norwich', his mother's house. I am much indebted to Dr M. Rajnai, of the Castle Museum, Norwich, for the above information.

Curtis had a sister living at 75 Welbeck Street, London, at least from December 1830–8. Whether she was married, a domestic servant or housekeeper or what her name was we do not know. As Curtis had letters and parcels directed there and as he eventually became a gentleman it is hardly likely that she was an ordinary servant, for he could hardly have used her employer's house as an accommodation address, and one doubts if a man with a sister a domestic servant could ever have become a gentleman. The 1841 census shows a Mr and Mrs Hay (John aged seventy-five, a confectioner, and Elizabeth aged fifty) and a Mr and Mrs Albert (Hugh aged sixty-seven, H. P. Army [*sic.*, probably H.M. Army], and Jane aged sixty) living at 75 Welbeck Street.

If Curtis's sister was Elizabeth Hay, aged fifty, the date of her birth would be 1791 and this would make her Curtis's twin sister. Whilst this is possible one feels it to be unlikely as one would have heard more about it: moreover the census return shows Mrs Hay to have been born in Middlesex. People do not necessarily declare their correct age in such returns. (Curtis himself falsified his age, reducing it by five years, in the 1851 census), but there are few who make themselves older than they are; on the other hand, Elizabeth, with a husband twenty-five years or more older than herself, might have added a few years to the return in order to make the discrepancy less; or, even more probable, Mrs Hay would be unlikely to have been consulted much about the census form and Mr Hay, the confectioner, would have filled it in according to his own lights and have increased (no doubt unbeknown to Elizabeth) the age of his spouse to avoid the reproach of cradle snatching. Alternatively Miss Curtis, if the housekeeper, might by chance have been away the night of the census, and thus not recorded at No. 75.

The rating returns show John Hay, the confectioner, to have been the occupier in 1832, paying a rent of £100 (and rates of £21. 13s. 4d.), making the suggestion that Mrs Hay was Curtis's sister slightly more probable and that John Curtis was asking his comparatively rich brother-in-law to receive his letters and parcels. It is all very speculative and is delving into yet another of the shadows that surround the Curtises.

Appendix 2: Latrielle

An interesting example of solidarity among entomologists may be seen in the case of the famous Latrielle, in welcome contrast to the various quarrels recorded in our text. Moreover, his life contains interesting parallels with that of John Curtis. Pierre-André Latrielle was born in 1762 in the Dordogne. Left an orphan at an early age he was protected by a local doctor and later by a prosperous merchant, who saw to his education. Latrielle, like John Curtis, was interested in natural history from an early age, but again, as later happened with the subject of this book, his protectors realized there was no career in this for a poor, though intelligent, boy and they directed his education towards the priesthood. Although he studied for it he did not enter the church and managed to eke out an existence in one way and another and by writing about insects, especially in their classification. Latrielle became suspect during the revolution, his name having been found as a novice or at least as training to be a priest. He was arrested, condemned, sent to Bordeaux and there imprisoned whilst awaiting execution. The prison doctor on his rounds one day noticed Latrielle carefully examining an insect and fell into conversation with him. Latrielle informed his visitor that it was a beetle – *Necrobia ruficollis* – that it was very rare and could he please send the specimen to two young naturalists in the town. It may be noted in passing that the insect is not particularly rare and that this was probably Latrielle's only way of letting his friends know where he was. Permission was given. The insect was delivered to MM. Durgelus and Bory de Saint-Vincent, who immediately made such a fuss at the imprisoning of one of the ornaments of France that Latrielle was released. A figure of the insect is engraved on Latrielle's tomb: he died in 1832.

Appendix 3: The Lepidoptera illustrated by Curtis and Stephens

The various possibilities suggested by these figures are set out below in non-statistical language, which may be painful to statisticians, but, I hope, more easily understood by that sound citizen, the man on the Clapham omnibus, so beloved by the lawyers.

There were 1,900 Lepidoptera to choose from. Curtis chose 199 and Stephens chose 119. Curtis preceded Stephens in each picture made. Only one coincidence was obtained, that is only once did they illustrate the same insect. Was this by chance or design?

There are four possibilities in looking at this matter:

1. Curtis and Stephens may not have cared whether they both did the same insects or not and have acted throughout independently of each other.
2. Stephens followed Curtis and, not wishing to do the same insects as Curtis did, would avoid doing any *already done* by Curtis.
3. Curtis would not do any insects already done by Stephens.
4. Neither Curtis or Stephens would illustrate insects done by the other person.

Let us now try and restate the problem in simple terms: Curtis and Stephens are sitting at a table with a bag containing counters numbered from 1 to 1900; there is a third person present, the arbitrator. Let us now consider the different cases.

Case 1. Both parties indifferent: Curtis picks out 199 counters, notes the numbers and puts them back, shaking the bag. Stephens picks out 119 counters leaving out the one he picks each time. How many counters is Stephens likely to pick that Curtis has already picked? The proportion of counters that Curtis has picked is 199 to 1900 or roughly 1 to 10. Therefore in every ten counters anyone subsequently picks out of the bag one is likely to be one Curtis had picked too. Stephens has 119 picks, so he is likely to select 119/10 or roughly 12 counters Curtis has picked. A disturbing factor is that the numbers of counters in the bag is dropping all the time Stephens is picking: on his last – 119th – pick there will be only 1781 counters in the bag of which 188 are likely to be Curtis counters. The proportion is not much affected by this. If they had been indifferent there would have been some twelve coincidences.

Case 2: Stephens avoiding Curtis selections: Curtis picks a number out of the bag, notes the number (say No. 24) and puts it back. Stephens then picks a counter. What are the chances that Stephens will then pick the same counter? Obviously 1 to 1900. If Stephens has not picked No. 24 the arbitrator removes No. 24 from the bag. Curtis now picks another counter (say No. 52), notes the number and puts it back. Stephens now picks a counter. What are the chances that Stephens will pick the same counter? Clearly 1 to 1900, minus 1, that is 1 to 1899. If Stephens has not picked No. 52 the arbitrator takes out No. 52 from the bag and Curtis makes another pick, notes the number, puts it back and offers the bag to Stephens and so on until Stephens has had 118 picks. At this point there will be 1702 counters in the bag. For the 119th turn Curtis picks out a counter (say No. 151), notes the number and puts it back and Stephens makes a pick. What are the chances he will pick the Curtis counter? Clearly 1 to 1782, substantially the same odds as when they started. If Stephens is avoiding doing any plates that Curtis had already done any coincidence is unlikely.

Cases 3 and 4 are substantially the same as Case 2, except that in Case 4 both the Stephens and the Curtis choices are removed from the bag and Stephens at the 119th pick would have been selecting 1 counter from 1664 in the bag, still long odds for a coincidence of choice.

The conclusion we can draw is that they were both avoiding doing insects already done by the other entomologist, otherwise they would both have done the same insect more often.

Appendix 4: Methods of Writing Dates

One of the minor differences of American and British practice is our respective methods of writing dates; one frequently finds that American practice is an old-fashioned British one, probably unchanged from colonial days. Americans tend to write dates month, day, year, and we write day, month, year, thus 1.12.72 is 12 January to an American and 1 December to an Englishman, two systems giving rise to endless confusion between us. It is interesting to note that Curtis and his contemporaries wrote their dates month, day, year (American style). Thus he was born on 9.3.91 to himself – 3 September, 1791.

Appendix 5: The Early Penny Post

Some surprising postal records were established when the volume of letters was so much smaller. Same day delivery in London was a commonplace. More surprising are the postmarks on Curtis's letter to Haliday mentioned on p. 66. On the face it has 'Hampsd. Road, 1 py. P. Paid' in brown ink and a diamond within which is '3m Feb 6 Q'; on the back within a circle is 'Belfast, Feb. 7, 1840'. Presumably this twenty-four hour delivery was a Roland Hill achievement. One doubts if it could be done today. Alternative explanations would be that the letter took a year and a day, or that the Belfast office had forgotten to alter its date stamp: neither seems likely.

Appendix 6: Economies of Pest Control

The pioneer farmers who take up a pest control process in its early stages are those who usually benefit most from it. In its later stages when all farmers are using the new method, the crop may be so greatly increased that the percentage drop in price due to abundance may be greater than the percentage increase in crop due to pest control; then the farmers as a whole are worse off, though the populace benefit from abundant and cheap food. So-called economic entomologists tend to ignore price-elasticity when putting a cash value on losses.

Losses can also be considered as an area of land lost, i.e. set aside for the purpose of feeding pests. For a discussion of these points see H. S. Smith *et al.* 1933[72], G. Ordish 1952[56] and G. Ordish and D. Dufour 1969[59]. The point was noted by John Curtis, see p. 89. Farmers generally benefit more from scarcity than abundance; see also Shakespeare. The porter in Macbeth cries (II, 3):

Knock, knock, knock! Who's there, i' the name of Beelezebub? Here's a farmer that hanged himself on the expectation of plenty.

Appendix 7:

One is reminded of Bernard Shaw's distinction on this theme. *Man and Superman*, Act III. The Sierra Nevada.

John Tanner and Henry Straker are led in by the brigands.

Mendoza: Allow me to introduce myself: Mendoza, President of the League of the Sierra! I am a brigand: I live by robbing the rich.

Tanner: (*promptly*) I am a gentleman, I live by robbing the poor. Shake hands.

Bibliography

1. *Archdeacon's Transcripts*, Norwich Records Office, 1785–1830.
2. J. Barbut. *Les genres des insectes de Linné* (London, 1781).
3. M. J. Berkeley. *Journal of the Royal Horticultural Society* (1846).
4. Berkenhout. *Synopsis of British Natural History* (3 vols.), (1769–72).
5. R. E. Blackwelder. *The dates and editions of Curtis's 'British Entomology'* (Smithsonian Miscellaneous Collection, 407. 5., 1947).
6. Robert Brown. 'Australian Insects', *Transactions of the Linnean Society*, **12** (1818).
7. E. Burton. *The Victorians at Home* (1831–61) (London, 1972), p. 95.
8. R. Burton. *The Anatomy of Melancholy*, memb. 2, subsect. 1.
9. Comte de Buffon. *Histoire naturelle* (44 vols), *vol. iv* on insects by Latrielle, *q.v.* (Paris, 1749–1804).
10. Rachael Carson. *Silent Spring* (Boston (Mass), 1961).
11. *Census* (No. 107. 684, 11 Robert Street, St Pancras, London), Public Records Office, London, 1841.
12. A. Cox-Johnson. *Handlist of painters, sculptors and architects associated with St Marylebone*, Local History Library (Marylebone, London, 1960).
13. J. Curtis. *British Entomology* (I, plate 143) (London, 1826).
14. J. Curtis. *A guide to the arragement of British Insects, being a catalogue of all the named species hitherto discovered in Great Britain and Ireland* (London, 1829 (2nd edn, 1837)).
15. J. Curtis. *Observations on the natural history and economy of the different insects affecting the turnip crop*, *Journal of the Royal Agricultural Society*, **II**(2), 193 (London, 1841).
16. J. Curtis. *Annals of Natural History* **5**, 110–21 (1850).
17. J. Curtis. *Farm Insects* (Glasgow, 1860).
18. J. Curtis. *Farm Insects* (2nd edn) (London, 1883).
19. J. Curtis. *Transactions of the Entomological Society*, 2nd Series, **II**, 19–21.
20. J. Curtis. *Transactions of the Linnean Society of London*, **XXI**, 39–42 (1852).
21. Baron Cuvier. *The Animal Kingdom* (London, 1834).
22. C. W. Dale. *The Lepidoptera of Dorset* (Dorchester, 1886).
23. C. Darwin. *The Origin of Species by Means of Natural Selection* (London, 1859).

24. Sir H. Davy. *Elements of Agricultural Chemistry* (London, 1813).
25. C. Dickens. *The Posthumous Papers of the Pickwick Club*, chapter 15 (London, 1836).
26. *Dictionary of National Biography*, Vol. VI, pp. 1054–66 (London, 1908).
27. *Dictionary of National Biography*, Vol. XVIII, pp. 1062–3 (London, 1909).
28. *Dictionary of National Biography* (1st supplement), Vol. XXII, pp. 524–5 (London, 1909).
29. C. Donovan. *British Insects* (10 vols.) (London, 1792–1801).
30. Duhamel du Monceau. *Histoire d'un insecte qui dévore les grains de l'Angoumois* (Paris, 1762).
31. Entomological Society, *Proceedings*, pp. 177–8 (London, 1863).
32. M. Faraday. *On new compounds of carbon and hydrogen, and on certain other products obtained during the decomposition of oil by heat*, Philosophical Transactions, 115, pp. 440–66 (London, 1825).
33. J. K. Galbraith. *The New Industrial State* (London, 1967).
34. T. W. Harris. *A Report on the Insects of Massachusetts Injurious to Vegetation* (Cambridge (Mass), 1841). (2nd edn. of above, edited by C. A. Flint, Boston (Mass), 1862).
35. *Hooperhoe papers*, (Letter from C. W. Dale, 21 Dec. 1901), Norwich Records Office No. 18331.
36. L. O. Howard. *A History of Applied Entomology, Smithsonian Miscellaneous Collection*, Vol. 84 (Washington, DC, 1930).
37. *Index to the Monumental Inscriptions in the Cathedral, Norwich*, Norwich Record Office (Rye 129), 1917, p. 103.
38. S. Johnson. *A Dictionary of the English Language* (Vol. II, 'Patron', 1765).
39. W. Kirby. *Monograpia Apium Angliae* (Ipswich, 1803).
40. W. Kirby. *Century of Insects* (London, 1817).
41. W. Kirby and W. Spence. *An Introduction to Entomology* (London, 1815–24).
42. 'A Lady', *Botanical dialogues between Hortensia and her four children* (London, 1787).
43. *Land and Window Tax Records*, Box 28 c (8) Precinct of Trinity Church, Norwich Records Office, (1789–96).
44. P. A. Latrielle. *Histoire naturelle générale et particulier des crustaces et des insectes* (4 vols.), Paris, An X (1799) and various editions (*see Buffon*).
45. P. A. Latrielle. *Considérations générales sur l'ordre naturelle des animaux composant les classes des Crustaces, des Arachnides et des Insectes* (Paris, 1810).
46. *Law Gazette: Stephens* v. *Rennie* (27 June 1833).
47. Linneaus (Linne, Carl) *Systema naturae* (Leiden, 1758).

48. J. Major. *A treatise on the insects most prevalent on fruit trees* (London, 1829).
49. S. Maunder. *The Biographical Treasury* p. 225 (London, 1859).
50. H. G. Millar. *Natural History* (London, 1795).
51. T. Mouffet. (trans. Topsell), *The Theatre of Insects*, p. 1009 (London, 1634).
52. A. Murray. 'On the Extirpation of Injurious Insects', *Journal of the Royal Society of Arts*, Vol. 25 (8 June 1877).
53. S. A. Neave. *et al. The History of the Entomological Society of London, 1833–1933*, pp. 91–2 (London, 1933).
54. S. A. Nebve, *et al.*, *The History of the Entomological Society of London*, p. 123 (1933).
55. E. Newman. *A familiar introduction to the history of insects, being a new and greatly improved edition of the grammar of entomology* (London, 1841).
56. G. Ordish. *Untaken Harvest*, pp. 16–18 (London, 1952).
57. G. Ordish. *Untaken Harvest*, chapter 3 (London, 1952).
58. G. Ordish. *Scientific pest control and the influence of John Curtis*, John Curtis Woodstock Lecture, 29 Nov. 1967, *Journal of the Royal Society of Arts*, **116**, pp. 298–311 (1968).
59. G. Ordish and D. Dufour. 'Economic bases for protection against plant diseases', *Annual Review of Phytopathology, Palo Alto. Cal.* **7**, pp. 31–50 (1969).
60. G. Ordish. *The Great Wine Blight* (London, 1972).
61. C. Pearl. *Victorian Patchwork*, p. 5 (London, 1972).
62. A. Pope. *Essay on Criticism: Collected Works*, Vol. I, p. 121, lines 1–4 (1785).
63. *Rating Book*. Holborn Central Library, No. 9223, p. 127 (London, 1841).
64. *Rating Book*. Marylebone Public Library, A.14 (1837).
65. R. A. Comte de Réaumur. *Mémoire pour servir à l'histoire des insectes* (4 vols.) (Paris, 1734).
66. Religious Tract Society, *The History of Insects* (2nd edn) (London, 1842).
67. J. Ross. *Appendix to The Narrative of a Second Voyage in Search of a North-West Passage*, p. lxi (London, 1835).
68. G. Samouelle. *The Entomologist's Useful Companion* (London, 1819).
69. W. Shakespeare. *Hamlet*, *iv*, 14–24.
70. G. Shaw. *General Zoology*, Vol. VI, Parts I and II (London, 1806).
71. A. Smee. *The potatoe plant: its uses and properties together with the cause of the present malady* (London, 1846).
72. H. S. Smith, E. O. Essig, H. S. Fawcett, G. M. Petersen, H. J. Quayle, R. E. Smith, and H. R. Tolley. 'The efficiency and economic effects of plant quarantine in California', *California Agric. Exp. Sta. Bull.*, **553** (1933).

73. W. Spence. *Tracts on Political Economy:* 'Britain independent of commerce', 1807. 'Agriculture the source of the wealth of Britain', 1808. 'The objections against the corn bill refuted', 1815. 'Speech on the East India trade', in vol. 1 (1822).
74. J. F. Stephens. *Illustrations of British Entomology*, Vol. I, p. 16 (London, 1828).
75. J. F. Stephens. *A Systematic Catalogue of British Insects* (London 1829).
76. J. F. Stephens. *Manual of British Coleoptera* (London, 1839).
77. J. F. Stephens. *Supplement to the Illustrations of British Entomology* p. iii (London, 1846).
78. J. C. Stevens. *A catalogue of the valuable library of entomological books of the late John Curtis Esq. . . . which will be sold by auction . . .* (Monday, 8th day of June 1863).
79. W. Swainson. *Zoological Illustrations* (London, 1820).
80. *The Times:* 19 July, 22 Nov., 24 Nov., 26 Nov., 1832, and 15 June 1833.
81. Virgil, *Georgics*, ii. 541. 2.
82. Priscilla Wakefield. *An Introduction to the Natural History and Classification of Insects* (London, 1816).
83. F. Walker. *Insecta Britannica, Diptera* (Vol. I, 1851; II, 1853; III, 1856, London).
84. R. Walker. *Flora Oxoniensis*, (The Flora of Oxfordshire) (Oxford, 1833).
85. R. Wallace. *Eleanor Ormerod, LL.D.* p. 56, p. 184 footnote (London, 1904).
86. I. O. Westwood. *Proceedings of the Linnean Society of London* 1862/3, pp. xxxv–xli.
87. I. O. Westwood. *Notice sur John Curtis*, Annales de la Société entomologique de France, 525–540 (1863).
88. Gilbert White. *The Natural History of Selbourne* (1789).
89. H. F. Wilson and M. H. Doner. *The historical development of insect classification*, St Louis, USA (duplicated) (1937).
90. Cecil Woodham-Smith. *The Great Hunger* (London, 1962).

Index